Shopping
for Health

ALSO BY SUZANNE HAVALA

Simple, Lowfat & Vegetarian

Shopping for Health

A Nutritionist's Aisle-by-Aisle Guide to Smart, Low-Fat Choices at the Supermarket

Suzanne Havala, M.S., R.D.

HarperPerennial
A Division of HarperCollinsPublishers

The information contained in this book is general and is not intended as personal medical advice for an individual's specific health problems. If you are seriously ill or on medication, please check with your medical doctor before changing your diet.

SHOPPING FOR HEALTH. Copyright © 1996 by Suzanne Havala. All rights reserved. Printed in the United States of America. No part of this book may be used or reproduced in any manner whatsoever without written permission except in the case of brief quotations embodied in critical articles and reviews. For information address HarperCollins Publishers, Inc., 10 East 53rd Street, New York, NY 10022.

HarperCollins books may be purchased for educational, business, or sales promotional use. For information, please write: Special Markets Department, HarperCollins Publishers, Inc., 10 East 53rd Street, New York, NY 10022.

FIRST EDITION

Designed by Alma Hochhauser Orenstein

Library of Congress Cataloging-in-Publication Data
Havala, Suzanne.
    Shopping for health : a nutritionist's aisle-by-aisle guide to smart, low-fat choices at the supermarket / Suzanne Havala. — 1st ed.
        p.   cm.
    Includes index.
    ISBN 0-06-095077-3
    1. Marketing (Home economics)—United States.
2. Low-fat foods—United States. 3. Vegetables in human nutrition. I. Title.
TX356.H38   1996
641.1—dc20   95-38389

96 97 98 99 00 ❖/OPM 10 9 8 7 6 5 4 3 2 1

To my parents

"A loaf of bread," the Walrus said,
"Is what we chiefly need:
Pepper and vinegar besides
Are very good indeed."

Lewis Carroll, *Through the Looking Glass*

Contents

Acknowledgments

I can only hope that other authors have as much fun writing their books as I have had writing this one. You may think that writing a book is a lonely ordeal. Perhaps you envision the author perched in front of a word processor late at night, bleary-eyed and surrounded by stacks of notes and crumpled pages of legal pad. (Throw in a couple of cat companions and late-night jazz on the radio, and you've got the right idea.) But this was only one part of the process for me. In fact, I was always connected with a network of family, friends, and colleagues who helped to make the experience such a rewarding one. I would like to thank each and every one of them for their support.

First, I would like to express my gratitude to superagent Patti Breitman, whom I greatly admire. This book was another of Patti's brainchildren; it also emerged, in embryonic form, out of earlier discussions with Charles Stahler of the Vegetarian Resource Group. I was delighted to have the privilege of working with Susan Friedland, my editor at HarperCollins, who is well known not only as an editor par excellence but as an author as well. The help of her assistant, Jennifer Griffin, was also greatly appreciated. Special thanks to Dean Ornish, M.D., who provided the foreword. I am thankful for his continued friendship and support.

The book manuscript was reviewed by Reed Mangels, Ph.D., R.D. (who braved this tome about food while at the pinnacle of queasiness early in her pregnancy); Kathleen Babich, B.S.N., R.N.; and Julie Hoskins, M.S., R.D., L.D.N. Their comments were invaluable, and I was very grateful for their input.

Numerous individuals generously gave their time to contribute anecdotes for this book. I was touched by their sincerity and genuine interest in sharing their experiences so that others could benefit. Many thanks to Cindy Blum, Berkeley Breathed and Jody Boyman, Patti Breitman, Peter Burwash, Joyce Goldstein, Jane Heimlich, Representative Andrew Jacobs Jr., Casey Kasem, Mollie Katzen, Deborah Madison, John McDougall, M.D., and Mary McDougall, Virginia Messina, M.P.H., R.D., Victoria Moran, Kevin Nealon, Cassandra Peterson, Robert Pritikin, Jennifer Raymond, John Robbins, Laurel Robertson, Martha Rose Shulman, Sally Silverstone, Debra Wasserman, Jane Wiedlin, and Spice Williams.

This book was researched at the beautiful Morrocroft Harris Teeter supermarket in Charlotte, North Carolina. The staff at Harris Teeter was kind and helpful and generously tolerated my loitering in the aisles for the better part of a year.

I would also like to acknowledge some of my friends and colleagues, not already mentioned, whose work I admire and whose support I have appreciated over the years, including Charles Attwood, M.D.; Neal Barnard, M.D.; T. Colin Campbell, Ph.D.; Hans Diehl, Dr.H.Sc., M.P.H.; Michael Klaper, M.D.; and Mark Messina, Ph.D.

In addition, the staff and volunteers at the Vegetarian Resource Group have been my family and friends for many years, including codirectors Charles Stahler and Debra Wasserman, nutrition advisers Mary Clifford, R.D., and Reed Mangels, Ph.D., R.D., medical advisers Jerome I. Marcus, M.D., and Arnold Alper, M.D., as well as Brad Scott, Rosanne Silverman, and Ziona Swigart. There are many others. I am deeply grateful and privileged to have the opportunity to be associated with this special group of people whose easy teamwork, intelligence, compassion, and dedication to public service are unequaled. I am keenly aware that such a wonderful experience may occur only once in a lifetime.

I am also grateful for the ongoing support of my friends at *Vegetarian Times* magazine and Cowles Media Company (Marianne Harkness, Stephen Lehman, Toni Apgar, Sharon Bloyd-Peshkin, and the crew), Worthington Foods, Inc. (David Schwantes and Frank Poston), and my friends and colleagues in Vegetarian Nutrition, a dietetic practice group of the American Dietetic Association.

I am lucky to be close to my family, an ever-present source of love, encouragement, support, and humor. I am grateful to my parents, Milt and Kay Babich; my sisters, Sandy Babich (and Sarah and Scott) and Julie Hoskins; and my brother, David Babich. My close friends, too, have been patient with me while I have been holed up like a hermit working on this book. I am thankful for their loyalty, understanding, and support.

This list is incomplete. However, in attempting to acknowledge those who have inspired and supported me, I am left with feelings of warmth, happiness, and gratitude for having these people in my life. I appreciate one and all.

Foreword

In a series of clinical trials conducted during the past eighteen years, my colleagues and I demonstrated, for the first time, that the course of even severe coronary heart disease often may be reversed in many patients by a program of comprehensive lifestyle changes, without coronary artery bypass surgery, angioplasty, or a lifetime of cholesterol-lowering drugs. These lifestyle changes include a very low-fat vegetarian diet, stress management techniques, moderate exercise, smoking cessation, and psychosocial support.

Rather than the usual course of getting progressively worse over time, patients in our study who made comprehensive lifestyle changes continually improved. These findings are giving many people new hope and new choices.

Within a few weeks, the patients in our research reported a 91 percent average reduction in the frequency of angina. Most of the patients became essentially pain-free, including those who had been unable to work or engage in daily activities due to severe chest pain. Within a month, we measured increased blood flow to the heart and improvements in the heart's ability to pump. And within a year, even severely blocked coronary arteries began to improve in 82 percent of the patients. LDL-cholesterol levels decreased by an average of 40 percent. Patients demonstrated even more reversal of heart disease after four years than after one year.

In contrast, control-group patients who made the more moderate changes in their diet and lifestyle recommended by many government agencies (i.e., eating less red meat, more fish and chicken, four eggs per week, and 30 percent of calories from fat) overall showed worsening of their coronary artery disease after one year. They became even worse after four years. This finding has now been confirmed in over a dozen randomized, controlled clinical trials.

We found that the primary determinant of improvement was adherence—in other words, the more people changed their lifestyle, the better they became. Surprisingly, we measured the most improvement in the oldest patient (now eighty years old) and in those who had the most severe disease.

Although our research has been focused on coronary heart disease, a low-fat vegetarian diet has many other benefits. For example, you can eat more and weigh less. We found that the average patient in our study lost twenty-five pounds during the first year, even though they were eating more food, and more frequently, than before. Also, a low-fat vegetarian diet is associated with reduced risks of some of the most common forms of cancer—for example, cancers of the breast, colon, prostate, lung, and ovary—as well as decreased incidence of high blood pressure, osteoporosis, diabetes, arthritis, and other degenerative diseases.

Why is a meat-based diet less healthful than a plant-based diet? Cholesterol is found only in animal products, which also tend to be high in both total fat and saturated fat, which your body converts into cholesterol. Meat is high in substances that oxidize, or convert, cholesterol into a form that makes it more likely to build up in your arteries: meat is low in antioxidants—beta-carotene, vitamin A, vitamin E, vitamin C—that help to prevent this buildup. Also, there is virtually no dietary fiber in meat.

In contrast, a low-fat vegetarian diet contains almost no cholesterol. Also, it is low in total fat, low in saturated fat, low in oxidants, high in antioxidants, and high in dietary fiber.

Making wise food choices is the first step in putting these dietary recommendations into practice. In *Shopping for Health,* Suzanne Havala shows how easy it can be to find what you need at your supermarket to eat a healthy, satisfying, low-fat vegetarian diet. There is an exciting array of foods from which to choose, and the choices are growing every day.

This way of eating creates feelings of abundance, not deprivation. *Shopping for Health* helps you choose wisely from the wonderful array of foods at your supermarket.

Dean Ornish, M.D.
President and director of the Preventive Medicine
Research Institute, Sausalito, California,
and author of *Dr. Dean Ornish's Program for Reversing
Heart Disease* and *Eat More, Weigh Less*

Preface

First, loosen your hold on any ideas you may have about diet books or books about nutrition (ho-hum). And free your mind of any hunches you may harbor about a supermarket shopping book written by a dietitian (this one doesn't wear a hair net or a lab coat). I won't tell you that you can't eat your favorite foods. I'm not the diet police. This is a book about choices.

Why, then, a book about grocery shopping? As a registered dietitian who has been active in the area of health promotion and nutrition education for over fifteen years, I have long been a secret observer of what people put into their grocery carts. Who hasn't noticed their fellow shoppers in the checkout line and silently evaluated the contents of their carts? Perhaps you've spotted an interesting new product in someone's cart and had the thought, *I always wondered who bought that.*

But what I also notice in the grocery store is the look of confusion and frustration on so many faces. I know that many people want very badly to make wise choices when they shop. They want to lose weight, to lower their cholesterol levels, to teach their children good eating habits through example, and they want to enjoy their meals. They've been hearing and reading about the latest dietary recommendations, but they are always left with the same question: What's left to eat?

Dietary recommendations are changing rapidly, and it can be hard to keep up. This week, beta-carotene is in the news, so we eat more carrots. But the next week, we hear that there are too many pesticides on the vegetables. Someone says we should take vitamin E to *prevent* heart disease, then we hear that too much iron can *cause* heart disease (but I thought iron was good for me?). Help! Not only does the news seem to change like the wind, but it's difficult to sort it all out, too. It's not hard to understand why many people have trouble when it comes down to the nitty-gritty of figuring out what to make for dinner tonight.

But before we go any further, a few comments about science and nutrition are in order. The nature of science is that it is always changing. As a result, nothing in nutrition is definite or carved in stone. What we know today will, no doubt, be improved on by future research.

While I was in college (I graduated in 1981), doctors considered cholesterol levels of up to three hundred milligrams per deciliter to be normal. Most hospital coronary care units served their patients eggs for breakfast. And I learned about the Pritikin diet and vegetarianism under the chapter on fad diets. We've come a long way since then.

Where are we now? Well, pardon the cliché, but fat is where it's at. The connection between a high-fat diet and diseases such as heart disease and cancer has become widely understood by the general public—so much so that dieters are now counting their fat grams, shoppers are eyeing food labels for the fat content, and food companies are creating new low-fat products and reformulating old ones.

Meat producers are calling attention to their efforts to raise leaner animals and lower-fat meats. Low-fat and nonfat dairy products are everywhere, while only several years ago consumers wouldn't touch them. Everywhere you look, "low fat" is the buzzword and "fat free" is the hook.

There's a problem, though, with focusing on just one aspect of the diet, such as the fat content. Concentrating on an isolated part of the diet—instead of on the big picture—can distract people from making more comprehensive changes in their diets, changes that can yield much bigger benefits.

Here's an example: people today are cutting down on the amount of fat they eat, and most of them do it by making substitutions such as skim milk for whole milk, lean meats for fattier cuts, or low-fat cheeses for regular, full-fat cheeses. Most of the fat in the typical American diet comes from animal products. So, in trying to reduce their fat intake, most people end up swapping one kind of animal product for another. The relative balance of animal to plant foods in the diet, though, remains essentially unchanged.

There is a growing contingency within the scientific community that is calling attention to this very point, however, and the recommendations they are making have become a megatrend in the field of nutrition. Data from the landmark China Project, which studied the eating habits and disease rates of people in the People's Republic of China, pointed to a diet that is very low in fat and essentially vegetarian as being optimal for human health. And headline-making research directed by Dr. Dean Ornish has shown that heart disease can often be reversed, without drugs or surgery, by making comprehensive lifestyle changes, including the switch to a very low-fat, vegetarian diet.

The direction that science has taken is creating

sweeping changes in government policy on food and nutrition and in the way Americans view food. The old four basic food groups has been retired, and the food guide pyramid has taken its place, emphasizing a diet based more on grains, fruits, and vegetables and less on meats and dairy products. (An even more progressive "new" four food groups model has been introduced by the Physicians Committee for Responsible Medicine. This model shows only plant foods—fruits, vegetables, grains, and legumes—as being the backbone of a healthy diet, with animal products being optional and limited.) And at the time of this writing, the United States Department of Agriculture is experimenting with changes in the federal school lunch program that will help lower the fat in children's meals and bring them more in line with current dietary recommendations.

So, as dietary recommendations continue to evolve, animal products are given an increasingly smaller role to play in our diets. Foods of plant origin have properties that tend to promote health and prevent disease. The more animal products our diets contain, the greater the health risks. And there may be even more compelling reasons to change the way we eat.

In his book *Eat More, Weigh Less,* Dr. Dean Ornish states, "I found that I feel much better when I don't eat meat, so for me it was a choice worth making even though I don't have any significant health problems. Since I began eating this way, I've never had to worry about my weight. I have more energy. I need less sleep. I think more clearly." For many people, what may be even more motivating than the health benefits is that they *feel* so much better when they eat well.

The extent to which you may want to change your diet is, of course, up to you. The greater the changes, the greater the benefits. But whether you want to take a step or two toward a healthier diet or whether you choose to go all the way and make comprehensive changes in the way you eat, you'll need to know how to navigate your way through the supermarket. How do I begin to apply the dietary recommendations that I'm hearing about? What can I fix for dinner tonight? What's left to eat? I hope this book will help you answer these questions and get started down the road to a healthier and happier lifestyle.

Suzanne Havala, M.S., R.D., L.D.N., F.A.D.A.
Charlotte, North Carolina

PART ONE

The Basics Before We Go

How to Use This Book

This is a book for people who want to improve their health and well-being by eating well. For many years, official nutrition recommendations have been less than direct and very conservative, in large part bowing to pressure from industry groups that stand to lose if people cut back on the use of their products. These watered-down dietary recommendations have failed to produce significant health benefits for most people.

That old saying, "everything in moderation," isn't very helpful when we're talking about the typical American diet. As Robert Pritikin, director of the Pritikin Longevity Centers, has stated, "in this country, we're *dying* of moderation."

In other words, the typical American diet is so extreme in fat and cholesterol content, that so-called moderate changes barely make a dent. The end result is still too much fat, too much cholesterol, and not enough fiber to offer significant health benefits.

This book will direct you to progressive dietary recommendations that call for radical changes in the way most Americans eat. These ideas are supported by an impressive body of scientific research and are being espoused by many of this country's leading health and nutrition experts.

For many years now, conservative dietary recommendations have called for taking in no more than 30 percent of the diet's total calories from fat. Scientists chose this level partly because they felt that it was a reasonable and attainable goal for Americans whose usual fat intake was far higher. Researchers were afraid that if the number was set lower, people would give up before they even started. So rather than setting the goal at a figure that was optimal for human health, the number was set at a level that scientists felt people would accept.

There is growing sentiment among the scientific community, however, that dietary recommendations for the public should reflect what is scientifically accurate, rather than what scientists think the public will accept. In other words, give people the unadulterated facts and let them make informed choices.

Today, there is abundant scientific evidence that an optimal diet for most people is much lower in fat and contains far greater amounts of plant material—fruits, grains, and vegetables—as compared to the more traditional diet that includes 30 percent of calories from fat. The healthiest diet is based largely on plant foods and has a fat content of 10 to 20 percent of calories.

This book will guide you to foods that fit easily into a plant-based diet that derives 10 to 20 percent of its calories from fat.

Those of you who are familiar with Dr. Dean Ornish's books, *Dr. Dean Ornish's Program for Reversing Heart Disease* and *Eat More, Weigh Less*, with Dr. Neal Barnard's *Food for Life*, or with the Pritikin or McDougall programs can use this book to help put into practice many of the principles that these authors set forth. Their programs advocate plant-based diets that derive 10 to 15 percent of calories from fat.*

Others can use this book simply to learn some of the pitfalls of supermarket shopping or to take the first steps in reducing fat intake and increasing the fruits, vegetables, and grains in one's diet.

This book takes you on a step-by-step tour of your neighborhood supermarket, with some side trips along the way. The core of the book provides an aisle-by-aisle nutritional comparison of products, along with discussion and recommendations. The focus is on what you can *have,* rather than on what you should avoid. Each segment includes a list of "great choices." Interspersed throughout the book are lots of hints and shopping strategies, along with examples of how some healthwise shoppers cope at the supermarket.

The book is divided into three parts. Part One includes this chapter's description of how the book is organized, which should help you determine how to use the book to your best advantage. The next chapter answers some basic nutrition questions that you may have and lays the foundation for the nutritional comparisons to come. In the final chapter in this section, I offer pointers on using the new food labels.

Part Two is the tour of the supermarket. I'll show you nutritional comparisons of foods and discuss the best choices. You should come away with a very clear picture of what to put into your basket. This is a realistic look at the supermarket, so foods are grouped just as they are really found in most stores.

For the most part, brand names are not used in

*These programs are very similar to each other. They share an advocacy of a very low-fat vegetarian diet. While this book is not designed as a companion to any one of these books, it is meant to help you make the transition to this type of diet.

this book. Occasionally, I mention a few products by name, especially if they are good choices. However, I decided not to fill the pages of this book with specific brand-name products for two reasons. First, new products are constantly being introduced into the stores. Most of them will fail and be discontinued; by the time this book is in print, many of the products I might have mentioned will be unavailable or will have been altered in some way. Second, products that are available in one part of the country are frequently not sold in other regions. Rather than give a list of products to which you may not have access—or list only those products that are available everywhere in the country—it made more sense to give guidelines and advice that would enable you to evaluate the foods that are offered in your area without relying on brand names.

Part Three is the glue that helps to hold everything together. It begins with a set of sample grocery lists. This section answers the second most frequently asked question (after What's left to eat?): Where do I begin? Just having a list in hand helps.

Next, we'll steal a peek into the baskets of some of the shoppers in the checkout line—a valuable form of people watching! I'll illustrate some sample grocery cart "makeovers," incorporating great choices that will improve the nutritional profile of the purchases.

The book wraps up with a list of books, cookbooks, periodicals, and organizations that are good resources for anyone who would like additional information.

Nutrition in a Nutshell

What should I eat, and will I get enough protein, calcium, iron, and so on?

Can I shop this way for my whole family?

How can I ensure that I'm eating well without getting bogged down with food diaries and calculators?

What's the Goal Here?

The goal is to move your diet toward one that is largely plant-based, greatly increasing the ratio of plant products to animal products in your diet. For most people, this means radically decreasing the amount of meat, poultry, fish, cheese, milk, and other dairy products that they eat. It also means cutting back on added fats, such as oils, salad dressings, and margarine, and replacing those fat calories with calories that come from low-fat, fiber-rich foods—vegetables, whole grains, legumes such as dried beans and peas, and fruits.

The dietary goal is fairly straightforward, but some of you may like more structure. A meal-planning guide is included in Appendix A toward the end of this book to help you see how the foods you eat fit into a healthful diet.

Another way of incorporating structure is to count fat grams. Many of you are familiar with this approach. More information about counting fat grams is provided in Appendix B.

However, most people should find that they do not need to rely on calculators, food scales, and measuring cups. By getting enough food to meet your energy needs—eating until full but not stuffed—and aiming for a reasonable amount of variety in your diet, the rest should fall into place.*

It's a Mind-set

Most of us have been conditioned to think of certain foods as being the primary, best, or only sources of certain nutrients. Meat, for instance, is associated with protein, and most people think of red meat when they are asked to identify a food that is rich in iron. How many people can name a food other than cheese or milk that is a good source of calcium?

We associate many animal products with being good sources of several essential nutrients. But the fact is, there are *many* foods that are good sources of these nutrients, and the greatest variety of these

---

*For those who are actively trying to reduce their weight, this approach works very well. However, at some point, calories *do* count. It *is* possible to consume enough calories—even from fat-free or low-fat foods—to prevent weight loss. This is most likely to occur when the diet contains a large amount of low-fiber, processed fat-free foods, such as fat-free cookies, cakes, and other snack foods. In contrast, fresh fruits, vegetables, and whole grains contain little, if any, fat, but they do contain large amounts of fiber. The bulkiness of these foods, as compared to fat-free, processed foods, helps to fill you up before you consume too many calories.

can be found in the plant world. The best foods are those that are high in health-supporting nutrients, such as vitamins, minerals, and fiber, and low in substances like saturated fat and cholesterol that contribute to disease when they are eaten in excess. And I'll explain in a moment that excess means a lot less than most of us are conditioned to believe.

A diet that is low in animal products protects against disease and promotes health. In its 1993 position paper on vegetarian diets, the American Dietetic Association states: "Studies of vegetarians indicate that they often have lower mortality rates from several chronic, degenerative diseases than do nonvegetarians."

However, when nutritionists recommend cutting back on meats and other animal products—or even eliminating them—the natural reaction from many people is "Then where will I get my protein (or my iron, my calcium, and so on)?" The concern about the nutritional deficiencies of a plant-based diet is mostly unfounded, as you will see. It is far riskier to continue to eat a traditional American diet that places heavy emphasis on foods of animal origin and promotes chronic, degenerative diseases and conditions such as heart disease, cancer, obesity, and diverticulosis, among others.

Good Foods Versus Bad Foods

A favorite tidbit of nutritional wisdom that has become the party line of many in the nutrition community goes something like this: "There are no good foods or bad foods. All foods can be a part of a balanced diet when moderation is used."

The idea is that we cannot assign moral qualities to foods and that there aren't any absolutes when it comes to prescribing how much or how little of a particular food a person should eat. It all boils down to balance. Eat a high-fat food now and balance it with a low-fat food later. Eat a plate of french fries and a grilled cheese sandwich for lunch, and balance that dose of fat with a nonfat supper of lentil soup and fresh fruit.

The problem is that the balancing in examples like this one still produces a diet that is too high in fat for most people. In reality, a truly healthy diet simply does not have room for most of the fat-laden foods that Americans love to eat. But certain food industry groups love this concept of balance and "no good foods or bad foods." It helps to keep dietary guidelines shrouded in ambiguity.

The concept of moderation doesn't work very well for most of us. I prefer a more direct approach. Let's be up front about how much is too much. Certainly, it's safe to say that certain foods are unhealthy in the amounts typically eaten by Americans, and it's also accurate to say that most

---

In 1992, I wrote an article for *Vegetarian Journal* titled "How Do You Define Moderation?" I interviewed many people for the article, including food writers, nutritionists, physicians, teens, consumer advocates, and others. Two of my favorite responses were the following:

"Regarding good foods/bad foods: There *are* some foods that are better than others. If you don't want to use the term 'bad foods'—then there are foods and there are good foods. The good foods you can eat anytime and the foods you can eat occasionally."

MARIAN BURROS, FOOD WRITER FOR THE *NEW YORK TIMES*

"I never use the term ['moderation'] because it's too vague and the food industry uses it as a smoke screen to make people think whatever they're eating is okay."

BONNIE LIEBMAN, DIRECTOR OF NUTRITION, CENTER FOR SCIENCE IN THE PUBLIC INTEREST

people would be no worse off—and would probably be better off—if they didn't eat certain foods.

Earlier, I noted that recommendations to limit fat intake to 30 percent of the calories in one's diet reflected what scientists thought people would be willing to do, rather than what they ought to do. The goal of 30 percent of calories from fat was considered moderation. Now we know that this brand of moderation doesn't produce significant health benefits for most people.

Again, this book will show you how to choose foods at the supermarket that fit easily into a health-supporting, plant-based diet that contains 10 to 20 percent of calories from fat. And whether you achieve this goal tomorrow or not, you can still use the book to learn how to make better choices and to take the first steps toward a healthier diet.

First and foremost, this is a book about choices. The point of view I take here is that you are entitled to accurate information on which you can base decisions about your own health and well-being—rather than being given only the information that someone else thinks you will want to hear. The extent to which you choose to change your diet is then up to you.

Nutshell Nutrition

A low-fat, plant-based diet can easily meet all known nutrient needs. Can deficiencies result? Sure, if your diet is comprised of soft drinks and fat-free coffee cake. A certain amount of care in planning any diet is necessary. But with reasonable planning, a low-fat, plant-based diet is more likely to provide all of the nutrients your body needs without providing the excess that accompanies the typical American diet.

Since most of us are conditioned to think of animal products as being the best (or only) sources of certain nutrients, it's natural to wonder where these nutrients will come from if the diet is plant-based instead. Here is a brief listing and explanation of the nutrients of significance and those about which people most commonly have questions:

PROTEIN

The body has a need for amino acids, which are the building blocks of proteins. All the amino acids needed by the body can be found in foods of plant origin; there is no human requirement for any animal products, such as meat, eggs, or milk from cows.

People used to think that if they ate a vegetarian, or plant-based, diet, then they had to combine certain foods with others in order to get a complete protein. (This idea was based on the myths that plant proteins are not as good as animal proteins and that plant proteins are unbalanced.) This is not true.*

Conscious combining of foods is unnecessary. Think of it this way: there are many, many interactions among nutrients taking place within your body every day. Just as you don't have to orchestrate these interactions consciously, you also don't have to combine protein sources consciously. Your body will put the pieces together by itself. (After all, if being adequately nourished required that kind of precision planning, how could any of us be alive today?)

There are two key points to remember on the subject of protein needs. The first: eat enough food to meet your energy needs. In other words, eat enough food to maintain a weight that is appropri-

---

* A good technical reference on this subject is V. Young and P. Pellett, "Plant Proteins in Relation to Human Protein and Amino Acid Nutrition." *American Journal of Clinical Nutrition* 59 (Supplement, 1994): 1203S–12S.

ate for you. The second point: eat a reasonable variety of foods, which can include vegetables, whole grains, legumes such as beans and peas, and fruits. If you do this, it is virtually impossible not to get enough protein.

One last point: an advantage to eating a plant-based diet is that while you are likely to get *enough* protein, you are also likely *not* to get excessive amounts. Foods of animal origin—meat, cheese, and eggs, for instance—are superconcentrated sources of protein, and these foods play a major role in the American diet. Too much protein is a problem for many, if not most, Americans.

Excess protein in the diet increases the body's loss of calcium by increasing the amount of calcium you excrete in your urine. Too much animal protein is associated with higher rates of some types of cancer and typically results in less fiber and more fat in the diet (since animal products usually contain fat and are devoid of fiber).

IRON

The form of iron that is present in animal products such as red meat is called heme iron. Heme iron is absorbed by the body more readily than nonheme iron, which is the form of iron found in plants. Nevertheless, people who eat a plant-based diet, including those who eat no animal products whatsoever, are no more likely to be iron deficient than those who eat more animal products.

There are various substances in our diets that can help the body absorb iron, and there are also substances that can inhibit iron absorption. For people who eat a reasonable variety of foods, these inhibitors and enhancers of iron absorption generally offset each other.

One of the substances that helps the body to absorb iron from plant sources is vitamin C, or ascorbic acid. There are lots of foods that are high in ascorbic acid—tomatoes, citrus fruits and juices, cabbage, broccoli, and green peppers, to name a few. So, eating a good food source of ascorbic acid with meals will help you absorb iron. Do you have a glass of orange juice with your cereal in the morning, or do you serve tomato sauce over your spaghetti? You probably already eat vitamin C–rich foods with your meals and don't even realize it. On the other hand, if you're a tea drinker, it may be wise to switch to herbal varieties or take up mineral water instead. Tea contains tannins, which can significantly inhibit your body's ability to absorb the iron in your diet.

In an ironic twist of science—and after years of people being concerned about getting *enough* iron—it has now been shown that excessive amounts of iron in the diet (and from supplements) are linked with increased rates of heart disease. For that reason, many health professionals are now advising people—particularly men—to avoid vitamin or mineral supplements that contain iron.

CALCIUM

The issue of calcium is a tricky one. In part, this is because the Recommended Dietary Allowance (RDA) for calcium is set at a much higher amount than is actually necessary, in order to compensate for calcium losses that are caused by the way most Americans eat.* The RDA for calcium is set so high that it's nearly impossible to meet it unless you take a supplement or eat foods that are superconcentrated in calcium, such as cow's milk.

---

*RDAs are the levels of intake of essential nutrients that the Food and Nutrition Board judges, on the basis of scientific research, to be adequate to meet the known nutrient needs of practically all healthy persons. For additional information, refer to National Research Council, *Recommended Dietary Allowances,* 10th ed. (Washington, D.C.: National Academy Press, 1989).

This brings us back to the protein issue again. As I mentioned earlier, too much protein in the diet causes your body to lose more calcium. If Americans ate less protein, they wouldn't need as much calcium. In fact, in many countries where the diet is more plant-based and protein intakes are lower, the recommendations for calcium intake are about half of those for the United States.

Because plant-based diets help to moderate protein intake—you get enough but not too much—your calcium needs are probably lower. Research has shown that no health risks are seen in vegetarians as a result of consuming less than the RDA for calcium—presumably because they absorb and retain more calcium than do nonvegetarians.

Calcium is widely available from many plant foods, and it's not hard to get what you need (unless you're a junk-food junkie). A few especially good sources include broccoli, bok choy, beans, dried figs, leafy greens (such as kale, mustard, turnip, and collard greens), and calcium-fortified cereals and juices. Since women, children, and teens have the highest calcium needs, they should be sure to eat these foods often.

VITAMIN B$_{12}$

This vitamin is produced by microorganisms that live on the surfaces of unwashed fruits and vegetables, as well as inside the gastrointestinal tracts of animals, including humans. The recommended dietary allowance for vitamin B$_{12}$ is incredibly small, and deficiencies take years to develop. All animal products—milk, meat, eggs, and so on—contain vitamin B$_{12}$. In our sanitation-aware society, fruits and vegetables are likely to be washed, so vitamin B$_{12}$–producing bacteria are rinsed away. Therefore, we can probably assume that plant products contain no vitamin B$_{12}$. When was the last time you pulled a carrot up out of your garden, brushed off the dirt, and munched away? Not in a long while, probably.

Vitamin B$_{12}$ isn't much of an issue unless you are someone who never ever eats anything of animal origin—no milk or milk products, no eggs. . . . In that case, it's best to be on the safe side and take a vitamin B$_{12}$ (cyanocobalamin) supplement, or eat vitamin B$_{12}$-fortified foods such as some commercial breakfast cereals or low-fat fortified soy milk.

VITAMIN D

Your body produces its own vitamin D when the skin is exposed to sunlight. There are some people, though, who may not receive regular exposure to the sun. Some very dark-skinned people, too, may have a particular problem making enough vitamin D if they live in a smog-filled city, in northern climates, or otherwise are exposed to little sunlight.

In this country, we protect such people by fortifying dairy products with vitamin D. If you are someone who does not use dairy products and you are also not exposed to sunshine regularly, it may be wise to check with your physician or a registered dietitian to see if a supplement is in order. Large doses of vitamin D can be dangerous, though, so don't exceed 100 percent of the Recommended Dietary Allowance.

Dodging Pesticides, Herbicides, and Other Environmental Contaminants

Many people are concerned about environmental contaminants—pesticides and herbicides—as well as additives, preservatives (more about these a little further on), and other unwanted ingredients in the foods they eat. While this book does not address these issues in great detail, a few words on the subject are in order.

I try to buy organically grown produce when it's available in order to minimize my exposure to environmental contaminants such as pesticides and herbicides. I weigh the practical consideration of cost—because organically grown produce that you buy in the supermarket is usually more expensive than other produce—against my desire for contaminant-free food. So when given the choice, I sometimes buy organic and sometimes not. If produce is not organically grown, then I wash it well (usually with a little dish soap and water) and peel anything that has a waxy coating on it.

While I am concerned about pesticides and herbicides (and buy organic when I can), I don't drive myself crazy trying to avoid every possible food contaminant.

One of the advantages of eating a plant-based diet is that plants contain only small amounts of environmental contaminants as compared to the amount found in animal products such as meats and dairy foods. Contaminants are concentrated in animal tissues, especially in the fat. Avoid the animal products and you avoid most of the problem.

A low-fat, plant-based diet is also high in fiber. A high-fiber diet decreases the amount of time it takes for food to work its way through the body and out. So if you have a high-fiber diet, contaminants will be less likely to be absorbed and will more speedily find their way out of your system. Fiber also binds with some contaminants and helps to remove them from your system. For these reasons, I don't get overly concerned when it isn't practical for me to buy organically grown produce.

"Pesticides are a red herring for the public. Worry about the bigger issues, such as fat—there's a bigger return for the effort. You can drive yourself nuts worrying about every aspect of your diet."

ROBERT PRITIKIN, DIRECTOR OF THE PRITIKIN LONGEVITY CENTERS, SANTA MONICA, CALIFORNIA

Additives, Preservatives, and Processed Foods

I advocate buying and eating foods that are as close to their natural state as possible. Foods in their natural state have no added sugar or salt, no artificial flavorings or colorings, and no preservatives. The more highly processed a food is, the more likely it is that nutrients have been lost and undesirable ingredients have been added.

So I like short ingredient lists on packaged foods. I generally avoid foods that contain certain additives that are not well tested or that have been shown to be of questionable safety. It bears repeating: the more highly processed a food is, the more likely it is to contain undesirable additives. Examples of these include artificial colorings, BHA and BHT (antioxidants used to prolong products' shelf lives), and saccharin.*

But that doesn't mean that you have to avoid *all* processed foods. Processed foods confer a degree of convenience. For many—if not most—of us, it is simply not feasible to forego all processed foods. Throughout this book, I have identified some pro-

*For more information about contaminants in foods, see Michael Jacobson, Lisa Lefferts, and Anne Witte Garland, *Safe Food: Eating Wisely in a Risky World* (Los Angeles: Center for Science in the Public Interest and Living Planet Press, 1991).

cessed foods that are acceptable on a low-fat, plant-based diet. In the cases where you may see some chemical-sounding ingredients listed on the package labels, those products contain no ingredients that are considered unsafe.

Convenience? Granted, some would argue that it doesn't take that much extra time to boil a fresh potato and mash it, for instance, as compared to whipping up some instant mashed potatoes. But for other people, even ten minutes saved is worth the trade-off. And for some, an occasional bottle of commercial, fat-free salad dressing is a necessary concession to convenience.

What is the cost of the convenience of processed foods?

Processed foods are more likely to contain chemical additives, and they may cost more than foods prepared from scratch. You do pay for convenience. Furthermore, some nutrients—vitamins, minerals, fiber—can be lost in the processing of foods (although some products are then enriched or fortified with vitamins and minerals). Also, many people contend that foods prepared from fresh, whole ingredients taste better than processed foods.

Whether or not you choose to include some processed foods in your diet is up to you. It will depend, in part, on the value you place on convenience, cost, taste, and nutrition considerations. You can, however, legitimately include some processed foods in a healthy diet.

Worry About the Dollars . . . (or Don't Sweat the Small Stuff)

You've heard the saying, "Worry about the dollars and not the pennies." Well, eating only fresh, whole, organically grown foods would certainly be ideal. However, it isn't realistic for most of us. We have to shop at supermarkets, and we lead busy lives that may leave little time for cooking from scratch. In order to eat the healthiest diet possible, you will have to prioritize various considerations about your food choices.

The easiest, most effective way to achieve the best diet is to focus on three key goals: 1) increase the ratio of plant to animal foods in your diet by centering meals around fruits, vegetables, whole grain breads and cereal products, and legumes—dried beans and peas—and decrease your intake of meat, eggs, and high-fat dairy products; 2) hold your fat intake to a minimum, and do away with added fats such as margarine, butter, regular salad dressings, and other added fats; and 3) limit your intake of empty-calorie sweets and other junk foods that don't give you much in the way of nutrition in return for the calories.

If you focus on these key points—rather than trying to juggle ten considerations—you'll get the biggest return for your effort while maintaining your sanity. When you put these ideas into practice, many of the other issues—concerns about pesticides, additives, and so forth—will also take care of themselves.

Can My Whole Family Eat This Way?

A low-fat, plant-based diet can be appropriate for everyone. Most experts agree that while adults may fare well on a diet that has 10 to 15 percent of its calories from fat, young children may need more fat in their diets in order to get enough calories to meet their energy needs.

In a family with young children, care should be

taken to give the children between-meal snacks and, perhaps, some higher-fat foods that the adults will want to avoid. Kids are growing and developing rapidly, and fat is a concentrated source of calories. Since young children have less stomach capacity than do adults, and since low-fat foods such as vegetables and fruits tend to be bulky, there is always the chance that a child may become full of these low-calorie foods before they have had enough to meet their energy needs. A little more fat in the diet can help provide those needed extra calories.

Most nutritionists recommend that children's diets contain about 30 percent of calories from fat, although that figure is certainly controversial. Others make a case for fat providing 20 to 25 percent of calories. I recommend aiming for a fat intake of 20 to 25 percent of calories for young children. But it should also be acknowledged that children in other parts of the world have fared very well on traditional diets that are much lower in fat than this. Much more research is needed on children's diets in the United States and Canada.

As with adults, it's better for children if extra fat calories come from plant sources and not from cholesterol-raising animal sources. With the exception of tropical oils such as palm or coconut oil (which are saturated fats and can raise blood cholesterol levels), it doesn't hurt for young children to get a little extra fat in their diets from such foods as avocados, seeds, nuts, peanut butter, and other nut butters.

And it still isn't necessary to wield a calculator in order to get it right. Nutritious snacks such as bean burritos, popcorn, bagels—even a sandwich or bowl of cereal—are great between-meal snacks for growing children. Follow your child's rate of growth on a height-and-weight chart (ask your pediatrician for a copy). You should see steady growth over the months and years. Only if there is a sudden drop in rate of growth should you be concerned. If that happens, a registered dietitian can help evaluate your child's diet for problem areas.

This brings me to one last word about children and low-fat vegetarian-style diets: people sometimes express fears that this kind of diet may lead to "failure to thrive"—that it might stunt a child's growth. These fears hinge on old attitudes and beliefs that a balanced diet must contain meat and dairy products and that if these foods are limited, then something vital will be missing. It also suggests that bigger is better. Once again, there are children around the world who have thrived for generations on a very low-fat, plant-based diet and reached full adult stature.* In the People's Republic of China, for instance, the average fat intake is less than 25 percent of calories; the diet is characteristically devoid of dairy products and contains little, if any, meat. The diet is extremely high in fiber. And the kids have no problems growing properly.

But the Chinese are also shorter than Westerners, you might point out. Unfortunately, there is very little research available on growth rates in Western children eating well-planned, low-fat vegetarian

---

*Studies that show "failure to thrive" in children eating low-fat vegetarian diets are few, but those that are cited were conducted on children in certain developing nations where the food supply, in general, was inadequate. The children did not have enough to eat. Or the studies were conducted on small groups of children eating overly restrictive or bizarre diets that were classified by the researchers as "vegetarian"—not reasonably planned diets such as the type described in this book.

In any case, it cannot be said that a child growing at the twenty-fifth percentile on the height-and-weight charts is any less healthy or more healthy than a child growing at the ninetieth percentile. Again, what is important is that the rate of growth is steady and that it does not suddenly fall off.

diets to use for comparison. Studies of American children eating lacto-ovo vegetarian diets, which exclude flesh foods but include eggs and dairy products and can contain as much fat as the typical American diet, generally show growth rates similar to those of other American children.

Vegan American children, who consume no animal products, typically consume less fat than other vegetarian children. There is little data available on growth rates of vegan American children, but there is no reason to believe that those eating well-planned diets should have problems. As a nutritionist who works with many vegetarian families, I am aware of many children who are thriving on vegan diets.

In its 1993 position paper on vegetarian diets, the American Dietetic Association states that "infants, children, and adolescents who consume well-planned vegetarian diets can generally meet all of their nutritional requirements for growth.*

There is no reason to believe that a well-planned, low-fat vegetarian diet is risky for children. On the contrary: considering that fatty streaks have been shown to develop in the arteries of even very young children eating the traditional American diet and that obesity is a major public health problem for our children, it's rather surprising that people aren't jumping up and down about the risks of kids *continuing* to eat a traditional American diet. The current system works as poorly for our children as it does for the adults in this country. Staying with current dietary practices is a far riskier proposition for everyone than switching to a prudent alternative.

*Position of the American Dietetic Association: Vegetarian diets. *Journal of the American Dietetic Association*, vol. 93, no. 11 (November), pp. 1317–1319.

A Label-Reading Primer

These days, you'll probably notice more people taking the time out in the supermarket to read the fine print on food products.

Thanks to new regulations that were put into effect in early 1994 by the Food and Drug Administration, food labels are now more helpful and easier to understand.

The new labels are designed to address the nutrition issues that are the most relevant to today's consumers. The labels focus on nutrients that Americans consume in excess—fat, saturated fat, and cholesterol, for instance—as well as on nutrients that many people need to consume in greater amounts, such as fiber, vitamins A and C, calcium, and iron (remember, vegetarians tend to get enough of all of these nutrients, assuming they aren't junk-food addicts).

The new labels are an improvement, but they aren't perfect. The most important point to note is that a fat intake of 30 percent of calories is used as a point of reference for some of the nutritional information that is presented for each product. In other words, the nutritional analyses that are printed on each label assume that a person is getting (or aiming for) a diet that has 30 percent of its calories coming from fat.

That's too much fat for most of us. As I mentioned earlier, the figure of 30 percent was an arbitrary number set because scientists felt that it was a goal that most Americans would not reject—not because it was the healthiest target.

Of course, some people only look at food labels in order to check the number of grams of fat per serving and to read the listing of ingredients. For those folks, the 30 percent glitch isn't much of a problem. But for those who want to eat a diet that is lower in fat and use the rest of the information provided on the label, you will have to make some simple adjustments. These adjustments will correspond to a diet that derives 15 percent of its calories from fat.

For instance, in the column labeled "% Daily Value," the value shown for total fat should be multiplied by a factor of two.* Look at the sample label that follows. In this example, one serving of this product is shown to provide three grams of fat, which translates into 5 percent of the Daily Value. To adjust this for a 15 percent fat diet, you need to multiply the number under "% Daily Value" by two. Using this example, one serving of this food actually provides 10 percent of your daily fat allotment.

The same method can be used to adjust the information given for saturated fat and cholesterol.

The reverse process can be used to adjust some of the information listed at the bottom of the label. Under the columns for 2,000 calories and 2,500 calories, *divide* the numbers shown for total fat, saturated fat, and cholesterol by a factor of two (or three, for a 10 percent fat diet). For example, under the 2,000 calories column, if the value for total fat

---

*For a 10 percent fat diet, multiply by a factor of three.

Nutrition Facts

Serving Size 1 cup (228g)

Servings Per Container 2

Amount Per Serving

Calories 90   Calories from Fat 30

	% Daily Value*
Total Fat 3g	5%
Saturated Fat 0g	0%
Cholesterol 5mg	2%
Sodium 280mg	12%
Total Carbohydrate 13g	4%
Dietary Fiber 6g	24%
Sugars 3g	
Protein 3g	

Vitamin A 80%	•	Vitamin C 60%
Calcium 4%	•	Iron 4%

* Percent Daily Values are based on a 2,000 calorie diet. Your daily values may be higher or lower depending on your calorie needs:

	Calories:	2,000	2,500
Total Fat	Less than	65g	80g
Sat Fat	Less Than	20g	25g
Cholesterol	Less Than	300mg	300mg
Sodium	Less Than	2,400mg	2,400mg
Total Carbohydrate		300g	375g
Dietary Fiber		25g	30g

Calories per gram:

Fat 9 • Carbohydrate 4 • Protein 4

shows 65 grams, dividing by two gives you about 33 grams, which is the amount you would aim for in a day on a 15 percent fat diet.

Of course, none of this calculating has to be necessary. Recognize that the new label format still supports the old dietary guidelines, which are tailored for a diet that is too liberal—too high in fat, too rich in foods of animal origin—to result in significant health benefits. The simplest way to use the new food labels may be just to focus on reading the ingredient lists while keeping an eye out for added fat and other unwanted ingredients, ensuring that the product is made mostly with whole grains, and so forth. Some people might also find it helpful to take a quick glance at the total fat content per serving, just to alert them if they missed some fats in reading the ingredient list.

Ingredients on food labels are listed in order of their predominance in the product. If an ingredient is listed as the first item in a food, then it probably comprises 50 percent or more of that product. For instance, a box of cereal that lists rolled oats as its first ingredient is made mostly of rolled oats. A bottled beverage that lists water, sugar, and fruit juice as its first three ingredients is mostly sugar water. You might compare that product to a similar one in which fruit juice is listed as the first ingredient, indicating that the item contains more juice than the first example.

One last note: at the time of this writing, food companies are in the midst of making the transition to the new label format. Thus, some of the foods on supermarket shelves are still wearing the old-format labels. Throughout this book, I have used real-life examples of foods, and sometimes the labels have presented nutrition information in the old format. In these cases, nutrient information may be listed as % USRDA instead of % DV.

The abbreviation % USRDA is short for Percent of the U.S. Recommended *Daily* Allowance. The Recommended *Dietary* Allowances are set at different levels for different age groups. The % USRDA, then, for a given nutrient was set at the level of the highest Recommended Dietary Allowance for all age categories.

In the new label system, a different approach is used. Daily Values (or DVs) are the new label reference numbers. They show how a food fits into a reference diet of 2,000 calories. (You may have to adjust figures up or down if you think your caloric needs are higher or lower than 2,000 calories.)

PART TWO

Let's
Go
Shopping

First Stop: The Deli/Bakery

Are you as hungry as I am? If so, then this place is dangerous! Supermarkets across the country may vary in their layouts, but most are similar in many ways. The deli/bakery area is very likely to be the first area you come to, and your cart isn't likely to be empty when you leave.

Supermarkets are devoting more and more space to deli and bakery items, featuring all sorts of prepared foods that are ready to eat or just need to be heated. Beautiful salads, sandwiches, fresh pizzas and pasta dishes, rolls, breads, and other bakery items pack this place.

Sure, this area is inhabited by doughnuts and pastries and Gouda and Brie. But there are some great low-fat and healthy choices tucked into every nook and cranny. And for (almost) every fat-laden coffee cake and fiberless gelatin mold, there's a much better choice just waiting to be discovered. Let's take a look at the choices here and see where the best ones are hiding.

Breads and Baked Goods

Scanning one side of this section, it's hard to miss the bakery cases. If you have a hankering for fudge brownies, pies, cakes, cookies, Danish pastry, doughnuts, cinnamon rolls, or sticky buns, you've arrived. If you had something else in mind, don't give up.

This is also where the muffins, bagels, and a huge variety of rolls and breads are found. But unique to this section is something you're *not* as likely to find: food labels on many of the baked goods. Some of the packaged breads have nutrition labeling on their plastic bread bags; others do not. Most of the bakery items have no nutrition labeling.

Not to worry. There are some ways to improvise and use whatever information is available in order to make choices. For instance, if packages don't list nutrition information, they may give an ingredient listing, and we can use that.

I've seen the ingredients for doughnuts and croissants listed on a card attached to the bakery shelf, and I've found bagel ingredients listed, along with the flavor of the bagel, on the signs on the bagel bins themselves. I've also found that store personnel can be very helpful. Ask if there is recipe information you can look at, or ask to see the box that the item arrived in, if it was baked off-site, and see if it lists the ingredients.

You've read the label-reading basics described in the previous chapter, so you know that the ingredi-

---

Talking shopping with Robert Pritikin, director of the Pritikin Longevity Centers, Santa Monica, California:

We have a sushi bar in my supermarket where I buy vegetarian sushi made with cucumbers. I also like to pick up a container of freshly squeezed carrot juice and some fat-free chips. That can be lunch sometimes.

ents on the label are listed in their order of predominance in the product. In the absence of nutrition information on labels, you can often make a fairly accurate estimation of the nutritional merit of an item by knowing what it's made of. Here's an example.

glazed doughnut	cinnamon raisin bagel
Ingredients: enriched bleached flour, water, partially hydrogenated soybean oil, high-fructose corn syrup	Ingredients: high-gluten flour, water, raisins, white sugar, malted barley flour, salt, yeast

No nutrition information for either of these two bakery items? Look at their ingredients. Both are made from flour and water, and both contain sugar. So far, they're similar. But the bagel also contains raisins—more raisins than sugar, judging by the order of the ingredients. Raisins are loaded with fiber and iron. Furthermore, the bagel is fat free. The bagel is a great choice. Better yet: look for a bagel made with whole grain flour.

A bagel can be a great substitute for a sweeter, fattier breakfast bread or pastry. Cinnamon raisin, apple cinnamon, and blueberry bagels are wonderful eaten plain—no need for cream cheese. If you need something on top, try a little jam or jelly; these condiments are fat free and much lower in calories than fatty cream cheese or butter.

Here's another example:

plain croissant	pita bread
Ingredients: unbleached enriched flour, water, butter, sugar, yeast, sweet dairy whey, wheat gluten	Ingredients: stone-ground whole wheat flour, water, yeast, calcium propionate to retard spoilage

Once again, in the absence of nutrition information, use your deductive skills to make the best

> **HELPFUL HINT**
>
> Shopping for muffins but find there's no nutrition information on the label? Here's another way to help you choose wisely. Just buy one. Take it home and set it on a napkin for a few minutes, then pick up the muffin. Check out the napkin. If there's a greasy spot where the muffin was sitting, you can be reasonably sure that it's very high in fat.

choice. Both the croissant and the pita bread are made primarily of flour and water. But the pita bread is made with whole grain flour, as compared to refined flour in the croissant. The third ingredient in the croissant is butter. But the pita bread is fat free—it's a great choice!

The deli at my favorite supermarket sells sandwiches made on croissants (some croissant sandwiches are also sold in the frozen foods section). Anyone who has ever eaten a croissant might have guessed that it was high in fat just by noticing the greasy film left on his or her fingers. Instead of making a sandwich with a fatty croissant, fill a pita pocket with chopped fresh vegetables, add some oil-free dressing, and you have a quick, easy, and healthy lunch or snack.

cheese bread

Ingredients: enriched bleached flour, water, cheddar cheese, yeast, high-fructose corn syrup, salt, hydrogenated soybean oil, whey

marble rye bread

Ingredients: enriched bleached flour, water, rye meal, rye flour, yeast, caraway, wheat gluten, salt, malt, partially hydrogenated soybean oil, sugar

Do you find you get tired of the same old loaf bread? You can build a much more interesting sandwich by using different breads. There are so many possibilities: onion rolls, pumpernickel and sourdough breads, light and dark rye, Kaiser rolls, French bread, crusty hard rolls, and the list goes on. With all of these choices, why bother with high-fat breads or rolls?

Actually, there are just as many *or more* low-fat, healthy breads to choose from as there are the fattier types. Compare the cheese bread and the marble bread. Flour and water are the first two ingredients in both. But the third ingredient in the cheese bread is high-fat cheddar cheese. (This bread is even greasy to the touch.) The rye bread is made, in part, with whole grain flour. And although it contains some fat, the amount it contains is very little. The oil is listed after the salt and the malt, both of which you might guess are present in very small amounts. So in the rye bread, the fat is a very minor ingredient.

The marble rye bread is a great choice. It's low in fat and made, in part, with whole grain flour. Besides, it's beautiful to look at, and it's delicious, too.

focaccia (black olive and tomato)

| Ingredients: flour, water, sliced black olives, tomato, olive oil, yeast, potato, salt | Serving size: 2 oz. (¼ of 8-inch round)
Calories: 150
Protein (gm): 5
Carbohydrate (gm): 23
Fat (gm): 4
Cholesterol (mg): 1
Iron (% USRDA): 4
Calcium (% USRDA): > 2 |

Italian bread shell

| Ingredients: enriched flour, water, part-skim mozzarella cheese, olive oil, yeast, salt, sugar | No nutrition information provided |

Here are two specialty breads that have recently arrived on the scene in supermarkets. Focaccia resembles a pizza crust with a light topping (usually a dusting of tomato and sliced olives, onion, or garlic), but without the tomato sauce and the cheese. It's typically just heated and eaten.

Another product, Boboli, is described as an "Italian bread shell." It also looks like a pizza crust with a very light topping of mozzarella cheese but without the tomato sauce you'd expect on pizza. Some people add typical pizza toppings and bake the bread like a pizza. It, too, can just be heated and eaten.

Judging by the ingredient listings, you might guess that these two breads are similar in fat content. Neither is made with a whole grain flour. The focaccia contains four grams of fat in a reasonable-sized serving—one slice. No nutrition information is given for the Italian bread shell, but the ingredient listings of the two breads can be compared.

Both are made primarily from white flour and water. The focaccia contains black olives, tomato, and olive oil, in that order. Black olives and olive oil are fat. The Italian bread shell is made with part-skim mozzarella cheese and olive oil, in that order. Although it's part skim, the mozzarella cheese still contains a fair amount of fat. And again, olive oil is all fat. The Italian bread shell probably contains as much fat, or a little more, than the focaccia.

How would I call this one? Neither bread is a nutritional powerhouse—they're low in fiber, moderate in fat content. I don't necessarily recommend either one. If you want to eat these breads, the best way to eat the focaccia or Italian bread shell would be to have one slice of either and round out the meal with other high-fiber, low-fat foods. Remember, as I mentioned earlier in this book, doing this kind of nutritional balancing act can work out if

---

HELPFUL HINT

For those who feel they have good control of their diets, it may be all right to incorporate some higher-fat foods into their meals once in a while. When you plan menus, you can balance higher-fat foods with foods that are very low in fat or fat free. For instance, if you are planning to eat a sandwich that is made with a high-fat bread such as a croissant or a fatty filling such as cheese (even so-called low-fat cheeses contain substantial amounts of fat) or hummus (a chickpea spread made with sesame butter), choose nonfat, fiber-rich foods to go with it—fresh fruit salad or maybe a crispy mixed green salad with lemon juice and pepper.

---

you otherwise have very good control of your diet. If not, then you should stick with breads that contain no added fat or only a minuscule amount.

Deli Prepared Foods

PIZZA PIZZA PIZZA

Who doesn't love it?! Supermarket delis are turning into pizzerias, selling ready-to-bake pizzas as well as fixing pizzas to order.

Pizza has lots of potential for being a delicious and healthful meal or snack. Unfortunately, it also has the potential to be packed with fat and cholesterol. The toppings are the key; the crust can also make a difference. Here's how to build a great pizza:

CRUST

The best crust is whole wheat, rather than a crust made with refined white flour. Whole wheat has more fiber and is packed with vitamins and minerals. Whether you have the option for a whole wheat crust or not, avoid deep-dish pizzas that have been slathered with shortening. (You know the type.) Regular pizza dough has a little oil added to it, but not much.

SAUCE

Most pizzas are made with marinara sauce—a meatless tomato sauce flavored with herbs and spices. There may be a little bit of oil added. Sauces that are meat flavored or have meat added are higher in fat. The best sauce is a plain marinara sauce. Some pizzas have no tomato sauce added. They're "white" pizzas, usually topped with cheese and assorted vegetables or meats, so they're high in fat.

CHEESE

Here's where the fat and cholesterol begin to come in. Some pizzas are made with regular, full-fat mozzarella cheese. Even part-skim mozzarella cheese contains a significant amount of fat, and if the cheese is sprinkled on with a heavy hand, the fat adds up. If you have the option of having a pizza made to order, try this: order the pizza without cheese. An alternative is to ask that only one quarter of the usual amount of cheese be put on the pizza, but consider this a stepping-stone and aim to eliminate the cheese eventually. You'd be surprised at how good a cheeseless pizza can be. *Really*—try it!

TOPPINGS

The best toppings come out of the soil—green peppers, onions, mushrooms, sun-dried tomatoes, pineapple chunks, slices of yellow squash or zucchini, artichoke hearts. . . . Olives are high in fat, so use them sparingly, if at all. The toppings to avoid are meats—sausage, pepperoni, ground meats, ham—as well as extra cheese!

Talking shopping with Kevin Nealon, comedian, _Saturday Night Live_, New York City:

(One of Kevin's favorite products is Pep-roni, a meatless pepperoni substitute)

I bring it to pizza restaurants, order a cheeseless pizza, and hand over my sliced Pep-roni. They're always happy to put it on!

You can apply the same advice to eating pizza as I offered for focaccia and the Italian bread shell: if you eat pizza made the conventional way—with some cheese or other fatty toppings—limit it to one piece with your meal. Then balance it out with low-fat, fiber-packed foods for the rest of the meal. Steamed vegetables, a baked potato, vegetable soup, mixed green salads, and fresh fruit are all great choices.

Talking shopping with Cindy Blum, opera singer, gourmet cook, and mother of two vegan children, Westminster, Maryland:

When I go to the local supermarket, I buy perishable items such as fresh fruits and vegetables. I judge a supermarket by the quality, diversity, and handling of its produce. I like to see organic produce offered in the supermarket, but it is not always the freshest or most reasonably priced.

I always read labels to find the products that meet my criteria for nutrition and/or environmental impact. I am a frequent asker of questions to store managers for products I would like but are not carried, and I also give frequent praise when I find something that is particularly wonderful.

Try new vegetables and fruits whenever possible. Most supermarkets have little cards describing "exotic" produce, often with serving suggestions and/or recipes.

FRESH PASTA, FRESH TOMATO SAUCE, FRESH SALSA

Fettuccine, linguine, spaghetti, angel hair . . . Some delis sell fresh pasta, but the packages usually don't contain nutrition information. They may contain a list of ingredients. If so, the list is likely to look like this: "extra-fancy durum flour and whole eggs." Pretty simple.

fettuccine

Ingredients: extra-fancy durum flour and whole eggs	Serving size: 3 oz.
	Calories: 260
	Protein (gm): 12
	Carbohydrate (gm): 45
	Fat (gm): 4
	Cholesterol (mg): 75
	Iron (% USRDA): 6
	Calcium (% USRDA): 2

The nutrition information in this list came from a commercial package of fresh pasta, which is typically found in the refrigerator or freezer section of the supermarket. As you can see, there is a small amount of fat in the pasta and a good-sized portion of cholesterol. The cholesterol—and most of the fat—come from the eggs.

A few stores that sell fresh pasta make an effort always to have on hand one variety that is made without eggs. Buy eggless pasta when you can. You might even want to use more dried pasta (see Aisle 4)—it's

---

HELPFUL HINT

Fresh marinara sauce (meatless tomato sauce) and fresh salsa from the deli are delicious. Use fresh marinara sauce and chopped fresh veggies to top a cheeseless pizza. Try using fresh salsa as a topper for baked potatoes, or use it in place of regular dressing on your favorite mixed green salad.

n available egg free and made with whole grain
r, too. Whichever type of pasta you choose, though,
it with a low-fat or fat-free sauce (like a good mari-
) and maybe some steamed fresh veggies.

PARED SALADS, SANDWICHES, SIDE DISHES, AND
ER PREPARED FOODS

This is only a partial list of prepared foods that
available in the deli section of one large North
olina supermarket:

- ole slaw (a few varieties)
- potato salad (a few varieties)
- una salad
- hicken salad
- eafood salad
- nacaroni salad
- arrot-raisin salad
- ntipasto salad
- our-bean salad
- pasta salad
- resh fruit salad
- green beans almondine
- baked beans
- esame noodles
- black bean salad
- orn relish
- grilled eggplant with tomato slices and
  mozzarella cheese
- pinach spaghetti with sun-dried tomato sauce
- basmati rice with apricots and pine nuts
- oasted new potatoes with rosemary

There is also a selection of ready-made sand-
hes, most made with meat and cheese, served on
ariety of rolls, bread, or croissants, as well as
ly-made tossed salads (most of which are topped
grated cheddar cheese and/or boiled eggs).

A couple of great choices pop out right aw
The fresh fruit salad looks wonderful. Made w
fresh seasonal fruit pieces and nothing else,
high in fiber, fat free, and full of vitamins and n
erals. Please note that buying prepared foods fr
the deli can be more expensive than if you bou
the ingredients separately and fixed the dish yo
self at home. But for people who live alone or
whom a whole batch of fruit salad would sp
before being eaten, it can make more sense to b
a single serving or two from the deli. With less fo
being wasted, the deli food can work out to be
more expensive than food made from scratch
home.

The baked beans also look like a great cho
Packaged in individual take-out containers, the
of ingredients reads: navy beans, ketchup, bro
sugar, onions, water, mustard, natural flavors, s
orange juice, soy sauce, wine vinegar, and garlic.

Vegetarians may wonder what "natural flavo
means.* But there doesn't appear to be any
added to these beans in the form of pork fat or
There's a little sugar and some other natural flav
ings, but the primary ingredient is supernutriti
navy beans. This is a great choice.

Another great choice is the four-bean salad, a
packaged in a single-serving take-out contai
While most bean salads contain added oil, this c
is oil free. It's flavored with vinegar and a l
sugar.

What about the other salads and prepared foc
The sesame noodles, black bean salad, and roas
new potatoes with rosemary look beautiful, as
many of the other dishes. Most of these foods
found in big bowls in the deli case, with no ing

---

* "Natural flavors" can sometimes include animal by-prod
Vegetarians may need to write or call the manufacturer to d
mine if this is the case in a particular product.

its listed. I asked a deli employee to let me see
recipes for some of the items in which I was
st interested.

Now, most everyone would agree that it's a good
a to eat more veggies, and pasta has been get-
lots of good publicity lately, too. All for good
on—vegetables, pasta, rice, and beans are all
mples of fiber-rich, nutritious foods. They're
ally relatively low in fat, too—that is, of course,
ess fat is added to them.

When I looked at the recipes for most of the deli
ds—salads made with pasta, rice, vegetables,
beans, all healthful ingredients—I was shocked
ow much fat had been added. A hefty dose of
e oil was added to every one. Even considering
each recipe made many servings, a quick anal-
using my calculator (yes, sometimes a calcula-
eally is necessary) determined that each serving
ained more fat than I would have ever guessed
far more than I wanted to eat.

ome people may reason that these salads are
er than the alternative—hot dogs and hambur-
, for instance. The salads still retain the benefits
eir nutritious ingredients—the fiber, vitamins,
minerals. And olive oil *is* better than animal fat
cholesterol.

a fact, olive oil has been the recipient of some
d press recently, where research has linked cul-
s eating a Mediterranean-style diet (which
es liberal use of olive oil) with lower rates of
t disease. However, it is likely that factors other
diet, such as exercise and genetics, play major
s in the lower rates of heart disease in Mediter-
an countries. Dietary recommendations in the
ed States stress the reduction of *all* dietary fats,
her from animal or plant sources.

he thing to be aware of is simply this: the deli
ls described, while otherwise full of healthful

ingredients, are high in fat due to the amount o
added to them. You may opt to eat them any
Others will decide to avoid them and choose lo
fat foods. I recommend the latter.

But without looking at the recipes, you m
not know that so much fat had been added. If
can put shyness aside, it can pay to ask store
sonnel to show you recipes or any other ingrec
information they may have for prepared foods
as these. Then at least you have the informa
you need and can make an educated choice. /
let store personnel know that you would like to
them offer more fat-free foods. Be assertive and
for what you need. If more people would
more options would be available.

Deli Meats and Cheeses

The bottom line first: if you use meats and chee
treat them as condiments—eat them sparingly.

HELPFUL HINTS

• When choosing deli salads such as cole sl
take an oil-and-vinegar variety instead of a may
naise-based salad (oil spreads more easily; it takes l
fat to cover the same amount of food). Even then,
the server to squeeze out as much of the dressing
possible when scooping your portion. Better yet, h
out for salads that are made with no added fats. Vi
gar, herbs, and spices are delicious and are all the
voring a salad really needs.

• Check deli recipes if you are unsure ab
whether or not (and how much) fat has been add
You may want to duplicate some deli salads at ho
Fix the same item but leave out some or all of the
meat, cheese, nuts, or other fatty ingredients.

ven low-fat and nonfat meats and dairy prod-
are a problem in the amounts consumed by
t Americans. Meats and dairy foods are concen-
d sources of protein—most of us get much too
h protein—and they contain no fiber. The more
t and dairy foods that are included in the diet,
more fiber-rich plant matter is pushed out or
laced. Plant matter—fruits, vegetables, and
ns—contains substances that promote health
protect us from disease. On the other hand, the
e animal products that are included in the diet,
greater the risk for chronic degenerative dis-
s. As the intake of meat and dairy foods
ases, so does the risk for heart disease, obesity,
blood pressure, cancer, diabetes, and other
ases.

ass up the deli meats and cheeses. It would be
difficult to include them in your diet in any
ficant way and still hold your fat intake to
it 15 percent of calories while keeping the
e of plant foods high enough.

LPFUL HINT

Clearly, this book advocates an extreme reduction
animal products in the diet. For those who choose
make this transition slowly, rather than making the
itch overnight, try this: make lean meats and low-
dairy products go further by getting into the habit
making them the minor ingredient—rather than the
in part—of a dish. For instance, use only one thin
e of either meat or cheese when making a sand-
h. Then load the sandwich up with slices of
nato, slivered cucumbers, spinach leaves, roasted
ppers, or other favorite vegetables. It's worth say-
again: use meat or cheese the way you would use
stard or pickles on a sandwich—as a condiment, if
all.

Supermarket Salad Bars

It's a sign of the times, but more people are ning to the corner supermarket to pick up lu Pay-by-the-pound salad bars are so convenie and not just for the lunch crowd. You may pay tle more, but sometimes it can make sense to together a quickie salad for two at the superma rather than washing lettuce and peeling carro home. Read more about fresh produce in the chapter. But a few pointers are in order for bars.

Load up on the fresh veggies—greens, toma green peppers, shredded carrots, beet slices, ar on. Then slow down when you hit the prep salads—macaroni salad, potato salad, pasta s and cole slaw, for instance. Some of these loaded with fat from oil or mayonnaise. I re mend avoiding them.

If you do take some, though, choose a with a dressing made of vinegar and oil instea mayonnaise (see the Helpful Hint earlier in chapter) and squeeze out as much of the dre as you can with the serving spoon. Let the dre on this item substitute for additional dressing you might have used on the salad.

Other fatty ingredients that are typically fou these supermarket salad bars are olives, seeds, and salad dressings. Skip them. You can add plain vinegar to the salad in lieu of the usual d ings if you are going to eat it right away. If yo taking the salad home, leave off the dressing a store and when you get home, add your fav flavored vinegar or salsa instead.

Some salad bars offer puddings. These are ally made with whole milk, so pass them by the other hand, fresh fruit is also sweet and ca a nice addition to a salad. Mandarin orange

---

ng shopping with Reed Mangels, registered dietitian
nutrition adviser for the Vegetarian Resource Group,
erst, Massachusetts:

tend to plan menus a week in advance, based in
on what's on sale producewise or what's in sea-
I shop with a list that is flexible (two vegetables,
les or pears, and so on). It helps to prepare a pot
oup or a casserole and use it for more than one
er. If you're pressed for time, use convenience
s like canned beans and quick-cooking grains. You
get precut vegetables from the salad bar for a
k stir-fry or salad on really rushed days.

general, look for short ingredient lists on pack-
s. Check fat content on labels. Read ingredient lists
ee if things like chicken flavor or lard have crept
seemingly vegetarian items.

---

ous, as well. Chunks of apple or pear are also
what unexpected and taste great. Raisins are
too.

is chapter has covered most of what you'll
n the deli/bakery area. But before leaving this
n, it would be hard to ignore the huge array
readable cheeses and dips that are found in
deli cheese cases. Here is the nutrition infor-
n provided for a few examples:

-name creamy spreadable cheese with garlic
ices

---

size: 1 oz. (about 2 T)
s: 100
(gm): 2
ydrate (gm): 2
): 9
erol (mg): N/A
n (% USRDA): 4
USRDA): > 2

brand-name light spreadable cheese with garden vegetables

Serving size: 1 oz.
Calories: 60
Protein (gm): 3
Carbohydrate (gm): 2
Fat (gm): 4.5
Cholesterol (mg): 15
Calcium (% USRDA): 2
Iron (% USRDA): > 2

avocado guacamole

Serving size: 1 oz.
Calories: 45
Protein (gm): 0
Carbohydrate (gm): 4
Fat (gm): 3.5
Cholesterol (mg): 0
Fiber (gm): 2
Calcium (% USRDA): > 2
Iron (% USRDA): > 2

None of these spreads or dips is low i
Even the "light" option has almost five grams
in a serving. Serving size is important to con
How far would one ounce (about two
spoons) of cheese dip or guacamole go?
many servings would you be likely to eat a
sitting?

Of the three choices shown here, two are
products. The other—the guacamole—is a
product, albeit a fatty one. Avocados are abc
percent fat. However, it's interesting to note
although it's high in fat, at least the guacam
cholesterol free and contains a little fiber-
grams per serving. The guacamole is the
choice of the three. However, all of them are
foods. Skip them all, and dip your crackers o
etable sticks in salsa instead.

The Last Bite: Bagel Chips, Bulk Candie
and Party Mixes

If bagels are a great choice, what about b.
chips? Let's evaluate a bag of them:

bagel chips

Ingredients: flour, water, partially hydrogenated soybean oil, sugar, corn flour, malt, salt, yeast	Serving size: 3/4 oz.
	Calories: 100
	Protein (gm): 3
	Carbohydrate (gm): 13
	Fat (gm): 4
	Cholesterol (mg): 0
	Fiber (gm): 1
	Calcium (% USRDA): > 2
	Iron (% USRDA): 4

There are four grams of fat in a serving. But
big is the serving? How much is three quarters o
ounce? Bagel chips are fairly heavy—much hea
than a potato chip. Three quarters of an ounce
small handful. In evaluating the merits of tl
chips, think about how many you realistic
would eat at one time and how much fat
would be.

We can also use the "finger test" in evalua
this one. Sample a bagel chip, or buy one bag a
experiment. After eating a chip or two, look at y
fingers. Is there a greasy film on them? Do your
gers feel oily? If so, then the bagel chips are pro
bly high in fat.

Bagel chips do tend to be greasy. A better ch
for a snack: eat a regular bagel or a few low
whole grain crackers. Or if you prefer, try ma
your own bagel chips. Slice day-old or stale ba
(be careful not to cut your fingers!) and spread
the pieces on a cookie sheet. Spray a thin film
vegetable oil spray over them, then sprinkle

c powder, paprika, oregano, cayenne pepper,
salt (if desired). Put them under a broiler until
just begin to turn brown. Then, turn the chips
a spatula and brown the other sides. Cool
re serving.

our store may also carry a variety of candy
ks in bulk bins. Some typical bulk candies
de yogurt-covered candies, milk chocolate–cov-
candies, butter mints, and assorted hard can-
. How do yogurt-covered candies, like malt balls,
ns, and nuts, stack up?

he yogurt covering on the candies is mostly
r and partially hydrogenated palm kernel oil.
urt cultures are at the bottom of the list. If this
urt-flavored coating covers nuts, then the candy
be very high in fat. Yogurt-covered raisins are
er in fat.

he only truly low-fat choice is the hard candy.
ight mints, sour balls, and other hard candies
probably the best bet, because they're low in
All of these candies are essentially empty-
rie foods—not much nutrition—with the excep-
of the raisins, which are nutritious but come
a dose of fatty yogurt covering.

pice drops, gum drops, and gummy candies are
low in fat. (Vegetarians may wish to avoid gum
os and gummy candies made with gelatin,
ough some are made with pectin instead. Read
abel or ingredient listing.)

arty mixes, such as tropical trail mix, Oriental
y mix, and others are also sold in bulk bins,
they're a mixed bag (pardon the pun). Some
es, such as the tropical trail mix, contain some
itious, high-fiber ingredients like raisins and
r dried fruits. On the other hand, they may also
ain fatty nuts and coconut. Other mixes usually
ain a variety of nuts and/or tiny cracker-type
edients that have been deep-fried or coated

with oil. Most of these mixes will leave a grea[se] film on your fingers.

Party mixes like these are easy to fix at hom[e] where you have more control over the ingredien[ts]. For example, you can easily mix together some [of] Chex-type cereals, raisins, and other dried fruits [to] make your own trail mix, leaving out the nuts a[nd] coconut. Or use dry cereals mixed with some c[ayenne] pepper, cayenne pepper, and other spices to creat[e a] party mix without the added oil. Or trade the pa[rty] mixes altogether for vegetable crudités with sal[sa,] fruit finger foods—grapes, berries—and fat-free p[op] corn for nutritious, high-fiber snacks.

GREAT CHOICES

- bagels: rye, poppy seed, sesame seed, oat bra[n,] onion, pumpernickel, plain, cinnamon raisin, app[le] raisin, blueberry, whole wheat, mixed grain
- rolls: garlic knot, whole wheat, onion, hard roll[s,] Kaiser rolls, pumpernickel, rye, hoagie rolls
- breads: pita pockets, French bread, Italian brea[d,] pumpernickel (light and dark), rye (light and dark[),] marble rye, sourdough, bread sticks, Hawaiia[n] bread, English muffins, multigrain bread
- angel food cake
- low-fat, whole grain muffins
- pizza (no cheese, using marinara sauce and veg[-] gie toppings)
- fresh marinara sauce
- fresh salsa
- sauerkraut
- pasta (made without eggs)
- gourmet fat-free salad dressings
- horseradish mustard
- fresh fruit salad
- baked beans (made without fat)
- four-bean salad (oil free)

alking shopping with Martha Rose Shulman, author of
everal excellent cookbooks, including *Fast Vegetarian
easts* and *Mediterranean Light*, Berkeley, California:

If you met me at the supermarket, here's what I
would have in my basket:

juice oranges (I squeeze them every morning so I'm
always buying them.)

bananas

broccoli (I usually buy broccoli when I shop, because
I know it will keep for a few days and I'll use it for
pastas and salads and just plain steamed for my
lunch.)

tomatoes, in season, for lunch, salads, pasta; canned
tomatoes otherwise for pasta

lettuce or field greens (I can't envision a meal without a salad.)

lemons

red bell pepper (I'll roast it if it begins to wrinkle
before I've used it some other way.)

nonfat yogurt

nonfat cottage cheese

dried pasta

canned chickpeas (if I don't have some on hand)

bread or unbleached and whole wheat flour for
making bread

Fresh Starts Begin Here: The Produce Section

From Asparagus to Zucchini

At last count, my neighborhood supermarket h
over one hundred varieties of fresh fruits and ve
etables from which to choose. Here is the one sp
in the supermarket where you'll find foods in th
natural state. Almost everything here is low in
and supernutritious. In fact, the whole area seer
to me to be screaming, "Fiber! Vitamins! Minera
Are you looking for beta-carotene or vitamin
This is the place to fill up your cart.

At the risk of sounding a little manic, I find t
produce area to be totally uplifting. Standing in t
middle, you are surrounded by beauty, color, a
fragrance. It's a little like enjoying the botanical g
dens or a floral shop. The next time you go sho
ping, plan to spend some extra time here.

While you're at it, take the time to walk slow
past each bin. Read each sign and take a look

ch of the items. How many of these foods have
u never noticed before?

When most people shop, they bypass about 70
rcent of the items in the produce area. They
ke a beeline for a few familiar items—the toma-
s, the iceberg lettuce, maybe some apples or
nanas. Does this sound like you? If so, it's time to
eak out of the rut.

The Produce-Area Challenge

few years ago, I had the pleasure of giving a
ies of supermarket tours to over a hundred peo-
living in the little town of Cornwall, Ontario—
of whom were adopting low-fat vegetarian diets.
ey were participants in the Coronary Health
provement Project, or CHIP, an excellent commu-
y health intervention project led by Dr. Hans
ehl of the Lifestyle Medicine Institute in Loma
da, California.*

Priscilla, a nurse visiting from the Caribbean
nd of Dominica, came along with me as an
server. The tour participants were in the produce
a when Priscilla captivated us all with stories
ut how some of these familiar foods are used in
native Dominica.

For instance, she explained that when she has a
nch of bananas that are becoming too ripe, she
els them and freezes them in plastic bags. Later,
takes them out of the freezer and uses them as
eded. Not only does she use them in baking but
Dominica, a typical way to eat bananas is to
sh them and spread them on toast or bread.
me cinnamon or nutmeg can be added. And

---

* At the time of this writing, CHIP has been offered in Kala-
oo, Michigan, as well as in Canada and India.

mashed banana can be stirred into a bowl of hot or cold cereal. Bananas are even delicious eaten cold straight from the freezer.

Did you know that bananas can be used as an egg replacer, too? (See the Helpful Hints that follow.) Priscilla renewed my enthusiasm about bananas. But there was more to it than that. She showed the tour group how to choose some of the exotic fruits that can be found in the supermarket today. She described how they taste and how they can be prepared. She introduced the group to items that they previously would have passed right by. Several of the tour participants picked up bananas and pomegranates that day!

So, here's the challenge to you: each time you go to the supermarket, take your time as you walk through the produce section and buy one item that you've never tried before. Or buy an item and use it in a different way than you have in the past.

You can figure that you'll find a few duds, but count on discovering many new favorite foods. I, too, eventually bought a pomegranate. I brought it home and cut out a wedge to try. Yes, the seeds were very beautiful. But the red juice stained my hands as well as my dish towel, and the taste was just so-so. I put the rest in the refrigerator, and it went bad before I ever ate it. (In defense of pomegranates, many people love them and would heartily disagree with me!)

I may not buy another pomegranate. But I've experimented with other foods over the years and have added lots of variety to my diet. I now crave steamed watercress and snack on kiwi fruit. I'm hooked on bags of prewashed, mixed Italian greens, and I've only recently discovered how wonderful fresh lime tastes squeezed over a baked yam. Have you tried broccoflower yet?

You Don't Have to Sacrifice Convenience, Either

's just plain hard to go wrong in the produce area. he list of great choices at the end of this section rovides just a beginning. Not only is there a sea of esh fruits and vegetables from which to choose ut many supermarkets are offering fresh conve- ience items as well.

For instance, jugs of freshly squeezed orange juice nd grapefruit juice are popping up everywhere. My eighborhood supermarket now sells fresh carrot ice and tangerine juice. Hate to wash salad greens? any supermarkets now carry bags of prewashed alad mixtures. Try the European mix of endive, dicchio, escarole, romaine, and iceberg lettuce. Or y my favorite, the Italian mix—baby lettuce, radic- io, and frisee. There are bags of shredded cabbage ixtures for cole slaw (mix it with an oil-free dress- g), and look for a fresh mixture of celery, onions, apa and Chinese cabbage for stir-fry.

For those who are concerned about preservatives n fresh veggies, most of these premixed greens dicate on the bags that they contain no preserva-

people mention the incidentals—the salad, the green beans, the potato? Chances are, not many. These are the side dishes, the items of less importance. The way a person responds to this simple question says a lot about where their mind is concerning the issue of food and how deeply entrenched they are in our traditional American ways of eating.

Begin to weed the menus that have meat as the main course out of your repertoire. Instead, stretch the meat in meals by using it as a minor ingredient and adding it to other foods. Instead of a baked chicken breast for dinner, make a big stir-fry with lots of fresh Chinese vegetables and add a handful of small chunks of chicken. Eat this over a plate of steamed rice.

But pause every now and then to reflect on how your eating habits are changing, and make sure that

---

HELPFUL HINTS

Instead of large portions of meat as the main course, try some of these dishes that use meat as a minor ingredient or condiment. Begin to use less and less meat as time goes by. Eventually, these dishes can be made meatless:

- Stir-fries made with lots of vegetables and tiny chunks of lean meat, chicken, or seafood, served over rice
- Multibean chili made with a small amount of lean ground meat
- Vegetable stew—a colorful medley of carrots, potatoes, assorted beans, tomatoes, and corn with small chunks of lean meat
- Pasta topped with sautéed vegetables and tiny bits of seafood in a marinara sauce
- Hearty lentil or bean soup with tiny bits of meat for flavor

you are steadily using less and less meat. Let your eating habits evolve to a point where you eventually stop using meat altogether.

Remember, too, that just as some people prefer a gradual approach to dietary changes, many others find that the whole process is less inconvenient when they make big changes all at once. They may have to muddle through the first few months, but they may also arrive at their goal faster than if they took the gradual approach.

Variety Is the Spice of Life

At my neighborhood supermarket, the produce section is situated adjacent to the meat counter. What a study in contrasts! The produce section is a feast for the senses—so many colors, shapes, textures, and delightful aromas. Items on the meat counter, on the other hand, are more similar to one another in color, texture, flavor, and use. It bears repeating: the variety really is in the plant world. Considering the sheer amount of meat that most Americans consume, most people actually eat a very limited number of different kinds of meat.

It's worth noting here that food companies are creating some delicious meatless alternatives to their real-meat counterparts. You don't necessarily have to go to a health food store to find them, either. Increasingly, manufacturers are buying shelf space in supermarkets for items like meatless veggie burgers, meatless hot dogs, meatless breakfast patties and bacon, and others.

In most cases, these meatless products are lower in fat than their real-meat counterparts, and they're usually cholesterol free, too. An increasing number of them are fat free. They're often located in the freezer case near the breakfast foods (you'd think they would be in the meat case, but they usually aren't).

Products like these are great transition foods. They serve a purpose for people who want to make the switch away from meat gradually. They can also be useful as convenience foods—to take along to a picnic or outdoor party or to keep at home in the freezer for an occasional quick meal.

For the biggest selection of these products, go to a large natural foods store. There are several companies that make delicious products, so you may want to experiment. At the regular supermarket, you may find one or two brands that are also sold in natural foods stores, and you'll also see products marketed by more mainstream companies. Read the labels and choose products that are fat free or very low in fat (three grams of fat per serving or less).

The Last Bite:
Planning Meals Without Meat

As a nutritionist specializing in vegetarian diets, I often hear the same remark: "If you don't eat meat, what *do* you eat?" Or to be more specific: "What do you make for dinner?"

Sometimes the easiest way to learn is to follow

Talking shopping with Cassandra Peterson, the actress who portrays the character Elvira, Mistress of the Dark, Hollywood, California:

I don't diet or do anything special with my diet, and yet I stay slim. I basically eat primarily vegetables, grains, and fruits, and I eat a *lot*. If you just cut back on or cut out meat, cheese, butter, and white sugar, then you can just about stuff yourself and not gain weight.

someone else's example. When I was very young, my mother decided to become a vegetarian. She never discussed her reasons for making the switch, and her diet was never a topic for discussion. Every day, she fixed "the usual" for dinner for the family but made a separate vegetarian dish for herself.

Then, several years later, one by one, the kids in my family all switched to a vegetarian diet. By the time I went away to college, I was a vegetarian. I had no problem figuring out what to eat for dinner at the dorm or, later, what to fix myself when I lived in my first apartment. Looking back, I can see how much I learned simply by observing the example set by someone else.

If you don't know a vegetarian family whom you can observe, see if there is a local vegetarian society where you live. Many vegetarian groups get together regularly for potluck dinners, where each person brings a vegetarian dish to share. You'll be amazed at the variety of dishes people bring and how delicious and beautiful the food can be. And if you ever wondered how a holiday meal could get by without meat, attend a vegetarian potluck at Thanksgiving time—you'll be amazed.

For more ideas, check out vegetarian restaurants, and see the cookbooks listed at the end of this book.

Aisle 1

Natural Foods
Health Foods
Special Diet

Remember when rice cakes and tofu could only be found at the health food store? Then, one year, rice cakes crept into the neighborhood supermarket. Incredibly, tofu followed. And there's been a parade of other alternative products staking claim to supermarket shelves ever since. Many natural foods brands have now gone mainstream. Aisle 1 is where we'll find these foods.

What's So Natural About Natural Foods?

Historically, only a small segment of the population was interested in the products that were sold in traditional natural foods stores, or health food stores. Since there wasn't enough public demand for these products, sales volumes were relatively low, and grocers couldn't justify allocating shelf space to them at the neighborhood supermarket. So these products found their homes in small health food stores, where prices were high.

What makes natural foods different from the usual supermarket brands? There is no legal definition for "natural." "Natural foods" is a catchall term that refers to products sold in health food stores or natural foods marts and usually produced by small companies. These products do tend to share some common features, however.

Frequently, natural foods are made without refined flours and sugars. Cereals, for instance, might be sweetened with fruit juice instead of refined sugar, and whole grains may be used instead of refined grains. There may also be a wider variety of grains used as ingredients, including such uncommon grains as quinoa, millet, amaranth, spelt, or kamut.

Some natural foods have no added salt. Typically, products are labeled as being made with organically grown ingredients. Many natural foods are vegetarian, and many have an ethnic theme, such as Middle Eastern falafel mixes, tabbouleh, or couscous. Many natural foods are marketed as being low in fat, while others are actually high in fat.

So, the term "natural food" can mean many things. You are likely to see more and more of these products—with brand names that may be new to

---

Talking shopping with Cassandra Peterson, the actress who portrays the character Elvira, Mistress of the Dark, Hollywood, California:

I actually don't shop in a regular neighborhood supermarket all that often, because I prefer to buy fresh organically grown foods. For that, I go to a natural foods store or a farmers' market. But when I do have to go to a regular supermarket, I can still usually find a small section of the store where natural foods brands are sold.

In addition to some of these natural foods products, I buy low-fat yogurt, organic produce if it's available, tofu, and some of the meat-substitute products such as veggie burger patties and others. I buy whole wheat bread and cereals. I buy soy milk, because I'm lactose intolerant. I buy the sugar-free, "fruit only" jams, Dijon mustard, and lots of pasta.

you—at your neighborhood supermarket as the demand for "healthy" products increases and more of these items find their way into mainstream stores.

Rice Cakes and Beyond

You may smirk and think Styrofoam when the topic turns to rice cakes, but if you haven't tried them lately, you're in for a pleasant surprise. They're still fat free (or close to it), so they're a great choice. But they also taste great. They come in new shapes and sizes and in many flavors. Popcorn cakes are another variation.

Plain rice cakes and their many-flavored cousins make great alternatives to most traditional snack chips, which are usually loaded with fat and salt. Instead of potato chips, corn chips, and assorted other greasy foods, try apple cinnamon, ranch style, caramel, or one of the many other tasty new rice cake products. Some of the flavored varieties are salty, but unless you need to avoid excess sodium for medical reasons, they make a much better choice than other fattier snack chips. (And they're flavorful enough that you don't need to use them as shovels for high-fat dips and spreads!) Read labels for fat content—just to be sure—but most contain no added fats and are essentially fat free.

Take a look at this comparison:

ranch-flavored rice cake

Ingredients: rice flour, brown rice syrup, buttermilk solids, canola oil, rice bran, sesame seeds, sea salt, whey, cheddar cheese, dehydrated onion and garlic, tomato powder, soy grits, butter solids, vinegar powder, citric acid, parsley, spices, natural flavorings, extractives of annatto	Serving size: 1 cake
	Calories: 35
	Protein (gm): 1
	Carbohydrate (gm): 7
	Fat (gm): > 1
	Cholesterol (mg): 0
	Dietary fiber (gm): unavailable
	Sodium (mg): 60
	Calcium (% USRDA): > 2
	Iron (% USRDA): > 2

traditional cheese puffs

Ingredients: enriched corn meal, vegetable oil, whey, cheddar cheese, salt, sour cream, artificial flavor, monosodium glutamate, lactic acid, artificial color, citric acid

Serving size: 1 oz. (about 21 pieces)
Calories: 150
Protein (gm): 2
Carbohydrate (gm): 16
Fat (gm): 9
Cholesterol (mg): 0
Dietary fiber (gm): > 1
Sodium (mg): 300
Calcium (% DV): 0
Iron (% DV): 0

Aside from containing less fat, the rice cakes also contain more wholesome ingredients, with no artificial flavors or colorings.

If you still have a hankering for some cheese puffs, though, try one of the new fat-free varieties that can be found in this section. Some are also made with whole grains. They're great choices, too. My supermarket also carries a potato chip alternative—take a look at the label and nutrition information here and compare them to traditional chips. Just remember not to load them up with a fatty dip.

low-fat chips

Ingredients: whole potatoes, whole grain brown rice, natural seasonings, canola oil, sea salt, rice flour, vinegar powder, citric acid, malic acid, natural flavors

Serving size: 8 chips
Calories: 50
Protein (gm): 2
Carbohydrate (gm): 11
Fat (gm): 0.5
Fiber (gm): 0.25
Calcium (% USRDA): > 2
Iron (% USRDA): > 2

traditional potato chips

Ingredients: potatoes, vegetable oil, salt	Serving size: 1 oz. (about 18 chips)
	Calories: 150
	Protein (gm): 2
	Carbohydrate (gm): 15
	Fat (gm): 10
	Cholesterol (mg): 0
	Calcium (% DV): 0
	Iron (% DV): 0

Crackers and Cookies

Fat-free or low-fat alternatives to the regular crackers and cookies found elsewhere in the supermarket can also be found in the natural foods section. As I mentioned earlier, many of these are sweetened with fruit juice instead of refined sugar and are made with whole grains rather than refined flours.

The value of fruit juice sweeteners versus refined sugar is debatable. But the lower fat content of many of these products, along with the addition of whole grains, can make many of these products great choices. Take a look at one comparison of a natural foods brand cracker and a similar product made by a mainstream company:

HELPFUL HINT

A useful product that you may find in the natural foods section of your supermarket is Ener-G Egg Replacer. This is a powdered egg-substitute product that is made from potato starch and tapioca flour. A teaspoon and a half, mixed with two tablespoons of water, substitutes for one egg in baked goods. This convenient product works well in many recipes, such as breads, cakes, muffins, pancakes, and others.

fat-free herb crackers

Ingredients: whole wheat flour, sprouted barley malt, honey, 15-fiber blend (organic wheat bran, organic oat bran, rye, soy, barley, and corn), concentrated grape juice, natural herb and vegetable blend, nonfat milk	Serving size: 5 crackers (about 1/2 oz.) Calories: 45 Protein (gm): 2 Carbohydrate (gm): 9 Fat (gm): 0 Cholesterol (mg): 0 Dietary fiber (gm): 2 Calcium (% USRDA): < 2 Iron (% USRDA): < 2

mainstream herb-flavored crackers

Ingredients: enriched flour, vegetable shortening, sugar, high-fructose corn syrup, salt, leavening, black pepper, poppy seed, natural and artificial flavors	Serving size: 4 crackers (about 1/2 oz.) Calories: 70 Protein (gm): 1 Carbohydrate (gm): 9 Fat (gm): 4 Cholesterol (mg): 0 Dietary fiber: N/A Calcium (% USRDA): < 2 Iron (% USRDA): 2

The first cracker is nutritionally superior. It's made with whole grains and has no added fat. The second cracker is made of the usual suspects: refined flour, fat, and sweeteners. With four grams of fat in four small crackers, it wouldn't take many to blow your fat budget for the day.

For anyone with a sweet tooth, it's worth trying some of the fat-free whole grain cookies, muffins, and even granola bars that are being produced by several natural foods companies. You may need to experiment with different brands to find those you prefer. Products sweetened with fruit juice instead of refined sugar may take some getting used to, but some people will like the flavor immediately.

Cereals and Toaster Pastries

Natural foods companies produce a wide variety of ready-to-eat breakfast cereals. Once again, most of these are sweetened with fruit juice and are made with whole grains, including such uncommon grains as amaranth, spelt, quinoa, kamut, and others. These are great choices. They're high in fiber, low in fat, and are made with wholesome ingredients. Whole grain hot cereals are smart choices, too. And if you're a fan of toaster pastries, try the new fruit-juice-sweetened, whole grain varieties.

Soups, Sauces, and Salsa

Hmmm. . . . Fat-free canned soups and chili made with organically grown ingredients—as well as the quick-style "cup-of-soup" varieties. . . . Pasta sauces made with organically grown tomatoes (ditto for the salsa). . . . How do these products rate?

The advantage to some of these products, as compared to mainstream brands, is the sheer variety of choices in these foods. Take soups, for instance. Sure, some mainstream companies now make fat-free varieties. But take a look—the choices are still somewhat limited. Also, some of these soups contain bacon, ham, or other meat flavorings

---

HELPFUL HINT

If you love granola but want to limit your fat intake, try one of the fat-free granolas that are available now. Mix some fat-free granola with another ready-to-eat cereal for a change of pace. Some people even enjoy mixing three or more different cereals together. Add some chopped dried cherries, dates, apricots, or currants, too.

---

that some people (including vegetarians and others) prefer to avoid.

Natural foods companies, on the other hand, make a huge selection of fat-free or very low-fat soups. Some are canned and ready to heat, and there is also a good variety of the cup-of-soup style (open the cup, add water, and heat in a microwave oven, or open the cup and mix with boiling water).

When it comes to pasta sauces, the main advantage to those on this aisle is that many of the natural foods brands are made with organically grown ingredients, and some contain no fat and no added salt. Individuals have to determine for themselves if it's worth paying a little extra for these features. Mainstream brands of pasta sauce typically do contain some oil, but this fat can be reduced by mixing the ready-made sauce with some plain tomato sauce (see Aisle 4).

Likewise, regardless of who makes it, salsa is always fat free. Varieties made with organically grown ingredients, like some natural foods brands, may cost a little extra. Shoppers have to decide for themselves if they value organically grown ingredients. Many people choose products made with organically grown ingredients as long as they don't have to pay too much extra for them. (See the discussion of pesticides and environmental contaminants in foods in Nutrition in a Nutshell.)

Mixes and Other Convenience Foods

Along with some of the delicious natural foods soups on the market, some of my favorite products are convenience foods. The only mixes I ever use are natural foods–brand products, because they are usually made with whole grains, and they don't contain undesirable additives, such as artificial flavorings or colorings (which are often made with chemicals that are not well tested or are unsafe—see the discussion of additives in Nutrition in a Nutshell). Here's an example. Take a look at the gingerbread mix listed here that is made by a natural foods company. Then compare the ingredients to a similar mix made by a mainstream manufacturer. No nutrition information is available for the natural foods product in this case, however. Food companies are only required to provide nutrition labeling if their product label makes a claim (such as "low fat," "high in fiber," and so on). In this case, no such claim was made.

natural foods gingerbread mix

Ingredients: whole grain stone-ground whole wheat flour, unbleached white flour, molasses, sugar, vegetable shortening, baking soda, salt, ginger, cinnamon

mainstream-brand gingerbread mix

Ingredients: enriched flour, sugar, vegetable shortening, dried molasses, defatted soy flour, leavening, dextrose, salt, propylene glycol monoesters, wheat starch, corn starch, natural flavors, mono- and diglycerides	Serving size: 1 piece Calories: 220 Protein (gm): 3 Carbohydrate (gm): 35 Fat (gm): 7 Cholesterol (mg): 30 Calcium (% USRDA): 2 Iron (% USRDA): 10

Both of these products are probably high in fat—the primary ingredients are flour, sugars, and shortening. But the natural foods–brand mix is made, in part, with whole grain flour. The mix calls for an egg and some milk to be added. I would modify the recipe by using an egg replacer (see Aisle 11).

I also recommend using skim milk instead of whole milk (or low-fat soy milk, which I use in place of cow's milk). Also, I serve a small piece of gingerbread topped with a heaping scoop of applesauce. By keeping the size of the gingerbread serving small and adding fruit to the plate, the total fat content of this dessert or snack is reduced. Besides, it's a delicious combination.

There are many other mixes that have some advantageous features—primarily their whole grain status and freedom from artificial flavors, colorings, and other undesirable additives. Some are fat free; others contain some fat. Examples of some of the mixes that you will find on this aisle include instant refried bean mixes, potatoes au gratin, macaroni and cheese, veggie burger mixes, vegetarian chili mixes, and many others.

From my experience, the items that contain no nutrition information tend to be those that are high in fat. Read the ingredient listings and see where the oil or other fat is placed—is it high on the list or further down? Remember that ingredients are listed in order of their predominance in the product. If oil or another fat is listed in first, second, or third place, that product is probably high in fat.

Even if fat is listed toward the bottom of the ingredient list, you may still want to check the nutrition information on the label (if it's provided) to make sure that the product does not contain too much fat. A gram or two per serving is generally fine.

Some people (especially those who want to keep fat to about 10 percent of the calories in their diets) may prefer to avoid products that contain *any* added fats, including oil, butter, cream, sour cream, hydrogenated and/or partially hydrogenated oils, mono- and diglycerides, and so forth.

Other people may decide that they will use some of the products described here as once-in-a-while foods, while still modifying the package instructions to cut down on any extra fat-containing ingredients.

If you choose this route, just be sure that you don't delude yourself about how often once-in-a-while occurs.

Many mixes call for oil or margarine to be added in the preparation of the product. Take this whole grain pilaf mix, for instance:

whole grain pilaf mix

Ingredients: millet, barley, buckwheat, dehydrated vegetables, miso powder, salt, dried yeast, herbs and spices

The instructions call for two tablespoons of butter or oil to be added. My recommendation: leave the fat out altogether. If the pilaf tastes a little dry, add some lemon juice. Or moisten it by serving it with a fat-free tomato sauce or topping it with sautéed vegetables.

Soy Milk and Soy Mayonnaise

Many supermarkets now sell soy milk in the natural foods section of the store. What advantage is there to soy milk over regular cow's milk?

Talking shopping with Casey Kasem, radio and TV host, Los Angeles, California:

In my grocery cart at the checkout line, I'd have organic fruits of all kinds—apples, peaches, nectarines, bananas, plums, melons, watermelon, cantaloupe—fresh vegetables (also organic)—broccoli, cauliflower, cucumbers, tomatoes, leaf lettuce, celery—all the good things that can be used in a salad or steamed or in soup.

Seven-grain rice, lentils, beans. Tofu hot dogs and veggie burgers. For snacks: whole grain pretzels. And a lot of organic carrots for my juicer because I drink a glass of carrot juice every morning.

Some people are lactose intolerant—they can't digest the milk sugar found in cow's milk. Soy milk can be a good alternative to cow's milk for these people, as well as for people who are allergic to cow's milk and for some strict vegetarians or those who fear drug residues and other contaminants in cow's milk.

Some soy milks are fortified with calcium and vitamin B_{12}. These are particularly useful for vegans, or strict vegetarians, who never use any animal products and want a reliable source of vitamin B_{12} in their diets. Others like to have the extra calcium if they are unsure about how much calcium they may be getting from nondairy sources (such as vegetables, legumes, and grains). If teens or children use soy milk that is not fortified with calcium in place of cow's milk, then they should be sure to include plenty of other good sources of calcium in their diet, such as green leafy vegetables, dried beans and peas, whole grains, and others. (See the resource list at the end of this book for additional information.)

All soy milk is high in protein, and some varieties are low in fat. Read labels and choose products that are lowest in fat. You can also dilute regular soy milk to cut the fat content. Some people even prefer the consistency of diluted soy milk to traditional soy milk.

Soy milk comes in different flavors, including plain, vanilla, and chocolate. Plain soy milk has a slight beany aftertaste to which some people may have to adjust. Personally, I don't mind the flavor of plain soy milk, but I usually use the vanilla flavor, which I love. Soy milk can be used any way that cow's milk is typically used—in recipes, on your cereal, or by the glass. Read labels, though, and buy a brand that is low in fat. (Also, please note that soy milk is not appropriate for babies as a substitute for commercial infant formulas or breast milk.)

Eggless mayonnaise or tofu mayonnaise is also

available in the natural foods section of supermarkets. Are these good substitutes for the real thing?

This is largely a matter of taste. Eggless and tofu mayonnaise are both cholesterol free, but they're not necessarily fat free. Be sure to read labels. Some mainstream brands of mayonnaise or salad dressing are also fat free and cholesterol free, and some are reduced fat. Take a look at this label from a jar of tofu mayonnaise:

tofu mayonnaise

Ingredients: silken tofu, canola oil, brown rice syrup, grain vinegar, prepared mustard, salt, apple cider vinegar, vegetable gums	Serving size: 1 tablespoon Calories: 35 Protein (gm): 0 Carbohydrate (gm): 1 Fat (gm): 3 Cholesterol (mg): 0 Calcium (% USRDA): > 2 Iron (% USRDA): > 2

Although this mayonnaise is cholesterol free, it still contains three grams of fat per tablespoon. Regular mayonnaise contains about eleven grams of fat per tablespoon, and other reduced-fat varieties contain about four grams of fat per tablespoon (it depends on the brand).

My suggestion: buy any brand you like, but go for those that are cholesterol and fat free or reduced fat. If you use reduced-fat mayonnaise, use it sparingly. Better yet, get out of the habit of using mayonnaise. Use mustard, salsa, or chutney on your sandwiches instead.

The Last Bite: Assorted Other Products

You'll find a variety of alternative products on Aisle 1. This aisle is also labeled "special diet" because some people who are following special, therapeutic diets (diabetic, gluten free, and others) may find useful products here. This is where sugar-free hard candies can be found, as well as some wheat-free and gluten-free products for people with allergies or intolerances.

Also spotted on this aisle were a variety of energy bars marketed as meal substitutes or as a quick energy boost for athletes. These products look like candy bars, but they're made with mostly unrefined ingredients.

The bottom line on these: they may be convenient as an occasional snack for someone—like an athlete—who is extremely active and needs a tremendous number of calories in his or her diet. But for most people, it is not a very satisfying meal substitute. For a snack, a piece of fruit, a bagel, or a bowl of cereal would serve the same

---

GREAT CHOICES

- rice cakes, many shapes, sizes, and flavors
- popcorn cakes
- other fat-free snacks: cheese puffs, potato chips, pretzels
- whole grain, fat-free crackers, cookies, muffins, and granola bars
- fat-free granola
- other whole grain, low-fat cereals
- whole grain, fat-free toaster pastries
- fat-free or low-fat soups
- fat-free pasta sauces
- salsa
- whole grain bread, muffin, and pancake mixes (modified to omit added fats)
- other whole grain mixes (modified to omit added fats)
- applesauce
- natural sodas (as a substitute for mainstream varieties)
- bottled fruit juice blends
- whole grain, eggless pasta

purpose for less money and fewer calories.

Some additional fat-free products on this aisle that are good choices include:

organic applesauce in a variety of fruit flavors
natural sodas that are free of caffeine, artificial
  flavors, and colorings
a variety of bottled organic fruit juice blends
whole grain, eggless dried pasta

Talking shopping with Jody Boyman, photographer, Seattle, Washington:

(Jody, who eats a strict vegetarian diet, is married to cartoonist Berkeley Breathed, creator of *Bloom County* and *Outland* cartoons. As you will note from what follows, Jody shops for many of their foods—not all of them low fat—at natural foods stores. However, more and more of the brands that she mentions here can now be found in the regular supermarket.)

I buy all the food in this household, and Berkeley eats what I cook (mostly!); here is a partial list of what you might find in my basket:

lots of organic fruits and veggies, especially arti-
  chokes
Tofu Pups [meatless hot dogs], especially the chili
  variety by Yves
Grainaissance Sesame-Garlic Mochi
Yves Veggie Pepperoni
Santa Cruz Natural Organic Apple/Cherry Sauce
White Wave Dairyless Yogurt
Vanilla Rice Dream—and lots of it!
Guiltless Gourmet Spicy Black Bean Dip
Worthington Canned Frichik
Imagine Foods Dairyless Chocolate Dream Pudding
organic quinoa, red lentils, and jasmine rice
pesto, pine nuts, and lots of sun-dried tomatoes
San Gennaro premade polenta

Aisle 2

This aisle is loaded with lots of fun, great choices. It's home to all sorts of specialty items from around the globe. And if you aren't excited about condiments right now, you will be after you see how many great choices are sitting right under your nose. Especially versatile and underutilized are the many varieties of mustard and vinegar (the subtitle of this section could read, "Fun with Mustards and Vinegars"). But first, let's examine the dressings and other salad paraphernalia that occupy about 25 percent of the space on this aisle.

Salad Fixin's

DRESSINGS

All that my grandmother ever put on her salad was her trademark vinegar and oil (she always got the mix just right). Maybe a little salt, too. That's it. And it was wonderful.

Since then, dressing a salad has become very complicated. Or at any rate, *dressings* have become more complex. There are certainly many, many varieties from which to choose. With names like Mango Mama, Dijon and horseradish vinaigrette, pesto Parmesan, and lime cilantro, who cares about the salad? You might just want to eat the salad dressing straight.

There's been a salad dressing explosion, and among the new members of the ranks are reduced-fat and nonfat varieties. From what I have seen, the range of flavors for these is not as extensive as for the regular varieties. But there are quite a few, starting with skinnier versions of the old standbys—ranch, French, Thousand Island, Italian, and so on. Are these great choices? It depends. Take a look at three sample dressings:

regular Thousand Island dressing

Ingredients: soybean oil, water, sugar, chopped pickle, vinegar, tomato paste, egg yolks, salt, contains less than 2 percent of propylene glycol alginate, mustard flour, dried onion, spice, calcium disodium EDTA as a preservative, natural flavor, artificial color	Serving size: 2 T Calories: 110 Protein (gm): 0 Carbohydrate (gm): 5 Fat (gm): 10 Cholesterol (mg): 10 Dietary fiber (gm): 0 Sodium (mg): 310 Calcium (% DV): 0 Iron (% DV): 0

reduced-fat Thousand Island dressing

Ingredients: water, sugar, liquid and partially hydrogenated soybean oil, vinegar, tomato paste, corn syrup, chopped pickle, salt, egg yolks, contains less than 2 percent of food starch—modified, with potassium sorbate and calcium disodium EDTA as a preservative, onion, xanthan gum, artificial color, spice, propylene glycol alginate, mustard flour, lemon juice concentrate, natural flavor	Serving size: 2 T Calories: 70 Protein (gm): 0 Fat (gm): 4 Carbohydrate (gm): 8 Cholesterol (mg): 5 Dietary fiber (gm): 0 Sodium (mg): 320 Calcium (% DV): 0 Iron (% DV): 0

fat-free Thousand Island dressing

Ingredients: water, corn syrup, sugar, vinegar, tomato paste, chopped pickle, cellulose gel, salt, contains less than 2 percent of xanthan gum, with potassium sorbate and calcium disodium EDTA as preservatives, dried onion, propylene glycol alginate, natural egg flavor, phosphoric acid, mustard flour, spice, lemon juice concentrate, artificial color, yellow 6, natural flavor, red 40, blue 1	Serving Size: 2 T Calories: 45 Protein (gm): 0 Fat (gm): 0 Carbohydrate (gm): 11 Cholesterol (mg): 0 Dietary fiber (gm): 1 Sodium (mg): 300 Calcium (% DV): 0 Iron (% DV): 0

If you take a look at the nutrition information on the labels first, you'll see that the fat content of these dressings decreases in a stepwise fashion from the regular to the fat-free variety. The regular dressing contains ten grams of fat in a serving. The reduced-fat, intermediate version of the dressing contains four grams of fat for the same size serving. One step further: the last dressing is fat free.

How significant is the amount of fat in these dressings, and is any of these dressings a great choice?

Imagine yourself at a salad bar, plate of salad in hand, about to ladle on some dressing. How much do you think you would use? One scoop? Two? How many tablespoons do you think are in a single scoop? Probably about four.

Thousand Island dressing is thick and creamy. It doesn't spread over a salad as easily as, say, Italian dressing. So you may tend to use a little more on your salad than if you were using a dressing with a thinner consistency.

Looking at the analysis of the three dressings shown here, the serving size is listed as two tablespoons. My guess is that most people would use at least that much on a typical salad. If the dressing is the regular variety, that adds ten grams of fat to the meal. Ten grams is a pretty significant amount of fat in most people's diets. For most people, the reduced-fat dressing is a much better bet, and the fat-free dressing is the best choice of all. If you are serious about holding your fat intake to 10 to 15 percent of calories, stick to fat-free dressings.

On the other hand, you may have noticed that all of these dressings contain some additives and preservatives. As I mentioned earlier in this book, the decision to avoid ingredients such as these is a personal one. I prefer to eat foods that are relatively unprocessed and as close to their natural state as possible. As I've said, I like short ingredient lists.

So why would I deem acceptable *any* product that contains additives and preservatives? First, some of the chemical-sounding ingredients in these commercial salad dressings are actually safe. Xanthan gum and EDTA, for instance, are fine. I do, however, object to artificial color and most artificial flavors. See the discussion of additives in Nutrition in a Nutshell.

As I explained early in this book, it is difficult to avoid all processed foods these days, and it isn't necessary to do so in order to eat a healthy diet. Focus on the big issues—more grains, fruits, and vegetables, less meat and high-fat dairy foods, less fat in general, and fewer sweets and fatty junk foods. If you do that, then many other features of a healthy diet will fall into place.

What are the alternatives to bottled salad dressings? There are plenty. My own experience has been that the tastiest salad dressings have been the simplest ones. One of my favorite dressings comes from the cookbook *Bean Banquets* by Patricia Gregory (Santa Barbara, Calif.: Woodbridge Press, 1984). It's the simple dressing that goes on the chickpea salad—made with lemon juice, wine vinegar, oil, hot pepper sauce, paprika, tarragon, and salt. I cut back on the oil or don't use any at all.

Two other simple recipes can be found in a cookbook by Debra Wasserman—*Simply Vegan* (Baltimore, Md.: Vegetarian Resource Group, 1991). Pick it up at most bookstores. Debra's Fruity Salad Dressing (four ingredients) and Tangerine Dressing (two ingredients) are quick and refreshing, and they include no fat whatsoever. They're great alternatives to commercial dressings. But if you don't want to fix even the simplest of recipes, try the suggestions below.

MAYO

Next to the bottled salad dressings on the shelf, you'll probably also find the mayonnaise and mayonnaise-type salad dressing—the white stuff. This product, too, now comes in regular, reduced-fat, and nonfat varieties (see also Aisle 1).

My recommendation: if you use mayonnaise, buy the fat-free variety. Otherwise, buy the reduced-fat variety and use it sparingly. Mustard, chutney, and some other condiments on this aisle are great alternatives to mayonnaise. We'll look at them in a minute. Also, I found a product called horseradish sauce on this aisle—a mixture of horseradish and mayonnaise. Don't confuse this with plain

---

HELPFUL HINT

These salad-dressing alternatives are great choices—and they're quick, easy, and delicious. Instead of bottled dressings, try:

- flavored vinegars (tarragon, raspberry, and others)
- your favorite salsa
- fresh lemon juice and freshly ground pepper
- freshly squeezed orange juice, lime juice, or a mixture

Talking shopping with Virginia Messina, registered dietitian, author of *The No-Cholesterol Barbecue Cookbook* and coauthor of *The Simple Soybean and Your Health*, Port Townsend, Washington:

There is a big space in my cart reserved for the wonderful condiments that make low-fat eating such an easy pleasure. I love good gourmet mustards (my latest find is a raspberry-flavored mustard) and special vinegars, including balsamic, herb-flavored, raspberry vinegar, and chili vinegar. I also stock up on sun-dried tomatoes, fresh ginger root, dried fruits, wonderful salsas, a good selection of nonfat salad dressings, and a variety of spices and herbs.

---

horseradish, which is fat free. (Be sure to read labels.)

BACON BITS, CROUTONS, AND OTHER TOPPINGS

Other salad toppers found on this aisle include bacon bits and croutons, as well as some salad accompaniments, such as bread sticks and flat breads.

Bacon bits are generally made from soy, with artificial flavor and coloring added. I personally never use them because of these additives. However, if you apply the reasoning that I outlined earlier, bacon bits are fine when used as a condiment—an occasional sprinkling on a salad—within an otherwise healthy diet. They contain about one gram of fat for every two to three teaspoons.

Croutons, too, are relatively low in fat. Some are actually fat free, and they're the best choice. However, unless you eat them by the handful, regular croutons sprinkled on a salad are not likely to pose a problem. Croutons range in fat content from about one to three grams of fat for approximately six croutons.

As for bread sticks and flat breads—these, too, tend to be low in fat. Some flat breads are actually fat free, and some are even made with whole grains. Flat breads are usually good choices.

Ketchup, Mustard, Pickles, and Vinegar

These condiments are so commonplace, they're easy to overlook. But they are fat free and extremely versatile. Granted, some of them pack a hefty dose of sodium, but a little bit of mustard can go a long way, and some reduced-sodium brands of ketchup are available. Besides, looking at the big picture, salt or sodium is less of a problem for most people than is dietary fat.

KETCHUP

I have a memory of my grandfather pouring ketchup on virtually everything he ate: into chicken soup, over his cabbage rolls, on his eggs.

My eating habits aren't anything like my grandfather's. But I do appreciate the versatility of his favorite condiment. Ketchup makes a great fat-free topping for baked or boiled potatoes. Mix it with a little salsa for a spicier version. (This mixture is actually available now in the supermarket—it's right next to the regular ketchup.) Ketchup goes well with some sandwiches, on eggless noodles, and on some vegetarian entrées such as lentil loaf (and, yes, veggie cabbage rolls).

Relatives of ketchup include many types of marinades (some low in fat, others high in fat) and barbecue sauces. These can be great on almost any roasted or baked vegetables, such as corn on the cob, vegetable kabobs, and baked beans. Read the labels to see what these products contain. Most are relatively low in fat, but some contain a fair amount of oil. If oil is listed near the top of the ingredient list, then pass that product by. Most of these products are a mixture of sweeteners, tomato sauce, vinegar, and spices.

MUSTARD

I mentioned earlier that mustard can be a great substitute for mayonnaise. Any kind of mustard can transform a sandwich and give it a whole new personality. A little scoop of mustard can also be used in many recipes—sauces, dressings, casseroles, you name it—to add some kick.

Now, I admit to liking old-fashioned, ballpark-style yellow mustard. My favorite summertime sandwich is fresh tomato slices on toasted whole wheat bread with some yellow mustard. But have you tried Dijon mustard with tarragon, lemon dill, bumpy beer (a local North Carolina variety!), or

---

HELPFUL HINT

Try this foolproof and easy recipe for home fries. Ketchup or a ketchup-salsa blend are great on these. (Note: this is a throw-together recipe—no measuring allowed!)

- Coat a baking sheet with a thin layer of vegetable oil spray and preheat oven to 350°F.
- Wash some white potatoes and cut them up into wedges. Use as many potatoes as you think you'll need to serve. (Leftovers are pretty good if reheated, too.) Put the potato wedges into a mixing bowl.
- Sprinkle potato wedges with several liberal shakes of garlic powder, cayenne pepper, paprika, and oregano. Add a little salt, if desired. Toss until potatoes are coated.
- Spread potatoes on baking sheet. Spray them with a thin layer of vegetable oil spray.
- Bake for thirty to forty minutes or until potatoes are soft.

spicy brown mustard? There is hot mustard, hot and sweet mustard, and many, many others. You might decide you want to become a connoisseur of mustards instead of wines. Just steer clear of the mustard-mixed-with-mayonnaise products.

PICKLES AND ASSORTED RELISHES

Pickles and pickle relish are fat free and can add lots of flavor to sandwiches and salads—or they can be nothing more than a tasty garnish on the side of the plate. They're condiments, and they're just fine. Similar condiments include chowchow (a relish made of cabbage, sweetener, onions, tomatoes, red pepper, vinegar, and salt—one bottle says it can be eaten with "beans or collards") and assorted other pickled vegetables such as sweet pepper rings, okra, bell peppers, string beans, and asparagus. The only cautionary note here concerns olives—they're mostly fat. Avoid them—or if you indulge, watch how many you eat.

VINEGAR

If you make a beeline for the apple cider vinegar when you shop and disregard all of those other bottles nearby, you're missing out. In addition to red wine vinegar, the all-purpose white vinegar, and cider vinegar, there's a whole world of flavored vinegars that is worth investigating.

My all-time favorite is balsamic vinegar. Balsamic vinegar is a rich, dark, sweet, flavorful vinegar made from wine grapes. It's frequently sold in what looks like a wine bottle—very elegant. In fact, a beautiful bottle of balsamic vinegar makes an impressive host or hostess gift.

Balsamic vinegar is absolutely delicious on salads all by itself. In restaurants, balsamic vinaigrette dressing mixes balsamic vinegar, oil, and herbs. But try the vinegar all by itself. You may notice that balsamic vinegar is relatively expensive compared to regular

> **Talking shopping with Robert Pritikin, director of the Pritikin Longevity Centers, Santa Monica, California:**
>
> Enjoying spices can make it much easier to eat a low-fat diet. Also, vinegar works well in place of salt in foods. Acids such as vinegar give you some salt punch. Have you ever tried hot-and-sour soup at a Chinese restaurant? It's the vinegar that gives it its kick.
>
> We use all different types of vinegars. Some people like red wine vinegar on salads, but I prefer white vinegar on mine. Malt vinegar on potatoes is dynamite. When we bake potatoes at home, if there are any leftover, we store them in the refrigerator. Then, we slice them and bake them on a nonstick pan and eat them with malt vinegar.

vinegars. But a tiny bit goes a long way. You'll probably use too much the first time you try it, and you'll find that you use less and less each time thereafter.

There are lots of other wonderful flavored vinegars to try. English-style malt vinegar is good for dipping chunks of French bread. Raspberry vinegar is tasty on salads. And some vinegars are flavored with herbs, such as tarragon or dill. Like the balsamic vinegar, some of these herbed vinegars even make great hostess gifts. Many of them are bottled beautifully, with a long piece of the herb floating gracefully inside the bottle.

International Foods

This aisle is a gastronomical trip around the world. There are lots of interesting specialty items to be found here, and many of them are great choices. The trick is to read ingredient lists and any nutrition information that might be available. Take the time to browse through this section thoroughly. You'll be surprised at how many of these products you never noticed before.

GOURMET CANDIES

No doubt about it—there are some excellent candy makers around the world. Many of the chocolates and other treats found here are packaged beautifully. Go for the fruit-flavored sours and the lovely tins of hard candies made from sugars and natural flavorings. The pectin jelly beans are a great choice, too.

For an occasional treat, these are fine. All of them are fat free; the pretty packages make them especially appealing. (See also Aisle 5.)

You may wonder, How can an item that has little nutritional value, such as hard candy or fruit preserves, be listed as a great choice? These foods aren't in the same league as beans or broccoli.

Generally, to qualify as a great choice, a food should be low in fat, low in cholesterol, and high in fiber and complex carbohydrates. Foods such as these fit in well with current dietary recommendations and should make up the backbone of a healthy diet.

However, within a category of foods—such as sweets, condiments, or desserts—some foods can be earmarked as being much better than other choices in that category. The great choices in these categories are always low in fat. Even so, they should be eaten in more limited amounts than foods that are less processed and more nutritious. This is a good example of the gray area in nutrition and illustrates the complexity of giving out dietary advice. It's hard, sometimes, to establish clear-cut answers or rules.

FRANCE AND GREAT BRITAIN

Other great treats are the beautiful jars of French preserves and conserves. These are fat free, and the conserves are usually a fruit-only product that contains no added sugar. You'll find some interesting mustards here, too. Imported French dried cherries are also good—use them in baked goods, in your breakfast cereal, or as a snack mixed with other dried fruits.

The British contribution on this aisle is an assortment of excellent teas, along with cookie accompaniments. Steer clear of the high-fat cookies (see Aisle 5). For more information about tea, see Aisle 4.

GERMANY

A glance at the German cake mixes here shows that shortening (fat) is a major ingredient, so bypass these. The same goes for the full-fat packaged cheeses you'll find on the shelves. However, there are some choices here that are quite good and might provide a nice change of pace. Look for hearty, whole grain dark rye bread and cans of sauerkraut made from white and red cabbage. There are some seasoned vinegars here, too.

ITALY

If you've ever thought about making your own pasta, this is the aisle where you can find pasta flour. For the less industrious, you'll find ready-made dried pasta here, too (buy egg-free varieties). Watch out for the dried, cheese-filled ravioli and tortellini, though. They're high in fat. You'll also

---

HELPFUL HINT

Have you ever eaten a veggie Reuben sandwich? Try this: on a slice of hearty rye bread, spread some fat-free Thousand Island dressing, add one slice of fat-free Swiss cheese, and pile on some sauerkraut. Add the second piece of bread to make a sandwich, then lightly grill the sandwich on each side, using a nonstick skillet and a little vegetable oil spray, if desired. Heat until the cheese softens (fat-free cheeses often don't melt very well) and the sandwich is hot.

find a variety of imported Italian pasta sauces. Read about pasta sauces in Aisle 4.

Italian specialties also include some vegetable relishes, or antipasto. Some of these vegetables are packed in oil; some are not. Check labels. Roasted peppers, for instance, are typically packed in oil. I found artichoke hearts packed in water and salt, so they were fat free. You'll find imported Italian olive oils here. Read more about oil in Aisle 7 (page 200).

INDIA

Indian food can seem mysterious to the uninitiated, but there are some items here that you'll definitely want to try. The first is basmati rice. This variety of rice is particularly fragrant and flavorful. It's low in fat and high in complex carbohydrates—it's a great choice. Try it with some curried vegetables. Cans of imported vegetable curry are available on this aisle, but they contain no nutrition information. Here's an example—how would you analyze the following item?

vegetable curry

Ingredients: water, carrot, zucchini, tomato, turnip, onion, vegetable oil, curry powder, graham flour, garlic, sugar, mustard seeds, modified corn flour

Water and vegetables appear to be the primary ingredients. Yet this item does contain some added fat. My guess is that this item is moderately high in fat. Is this a good choice?

My recommendation is always to look for a fat-free alternative first. However, if you decide you want to try this product, then eat this item over a heaping helping of basmati rice. This would be a delicious, nutritious meal, and the fat content would be fairly low overall if the product was stretched by eating it with lots of low-fat rice.

Incidentally, have you ever wondered, What exactly *is* curry powder? Curry powder is a mixture of several pungent ground spices. Dishes that are seasoned with curry powder are sometimes called "curries." For instance, cooked vegetables that are seasoned with curry powder might be referred to as a "vegetable curry," as in the example just given.

Curry has a unique aroma and adds a distinctive flavor to Indian dishes. Its spices may include cayenne, fenugreek, turmeric, cumin, garlic, coriander, ginger, cardamom, nutmeg, paprika, cinnamon, and cloves.

Another terrific Indian specialty item is chutney. Chutney is a condiment made with fruits, usually including raisins or dates, onions, and spices. A jar of traditional mango chutney bears no nutrition information, but the list of ingredients includes mangoes, corn syrup, sugar, distilled vinegar, salt, raisins, lime juice, dextrose, tamarind extract, caramel coloring, spices, natural flavors, and dehydrated onions.

Chutney is a great choice. Try adding it to a sandwich. Ripe tomato slices on whole wheat toast (with or without one slice of low-fat cheese) with a scoop of chutney is delicious.

You may also see some boxed mixes for various kinds of Indian breads, such as chapati, nan, papadum, and others. Read the instructions and ingredient lists carefully. Some of these call for fat to be added or for the batter to be deep-fried. Not all of them provide nutrition information. How would you evaluate the following bread mix?

nan mix

Ingredients: wheat flour, vegetable fat, dried yeast, salt, lactic and citric acids	Serving size: 100 gm
	Calories: 379
	Protein (gm): 11
	Carbohydrate (gm): 67
	Fat (gm): 6.7
	Dietary fiber (gm): 3.8

One of the first things you might notice about the nutrition labeling on this item is that the serving size is listed in grams rather than "slices of bread" or number of servings. A slice of loaf bread weighs about twenty-five grams. If you know that, then you might have decided to divide the information given here by four in order to evaluate it. In that case, you'd see that one serving would contain a little less than two grams of fat—not much.

An alternative approach is to determine what percent of calories in this food comes from fat. This method can seem pretty complicated, and I don't know many people who like to compute their purchases on a calculator while they're shopping. Don't be put off; these calculations will help you.

There are nine calories in one gram of fat. The nan mix above has 6.7 (round to 7.0) grams of fat, so about 63 calories come from fat (7 grams times 9 calories per gram). Sixty-three of the 379 total calories come from fat. Therefore, 63 divided by 379 equals 0.166. Multiply this number by 100 to get the percentage. Thus, 16.6 percent, or about 17 percent, of the calories comes from fat.

A piece of run-of-the-mill whole grain loaf bread derives about 22 or 23 percent of its calories from fat. A typical piece of loaf bread, for instance, contains about two grams of fat—not much—and about eighty calories per slice. Using the information provided here, then: 2 grams of fat per slice multiplied by 9 calories per gram of fat equals 18 calories from fat.

If there are 80 calories in a slice, then 18 calories divided by 80 equals .225 (round to .23) or about 23 percent of calories from fat.

So, the piece of nan in the example earlier has a little less fat than a typical piece of loaf bread, and that's acceptable for a healthy, low-fat diet.

You might have guessed that the nan mix was higher in fat, since vegetable fat is listed as the second ingredient on the box. But because the fat content is actually so low, the first ingredient—the wheat flour—must make up the vast majority of this product. Nan is a low-fat Indian bread.

GREECE

Greek *glykadi* is billed as the Greek challenge to Italy's balsamic vinegar. Give it a try. In this section, you'll find imported Greek dolmas—grape leaves stuffed with rice and nuts and packed in oil. If you'd like to try your hand at making a lower-fat version at home, pick up some of the grape leaves packed in brine and stuff your own with a rice and herb mixture.

You'll also find some pickled vegetables, such as pickled peppers, in this section. These are fine. You may also see jars or cans of tahini, which is a sesame seed paste—similar to natural-style peanut butter, with the oil rising to the top. Like peanut butter, this is mostly fat, and you should avoid it or use it only very sparingly.

Middle Eastern hummus is a spread made with garbanzo beans, lemon juice, white pepper, garlic, olive oil, and fat-laden tahini. I like to make my own version in a blender at home, adding a little more lemon juice, skipping the oil, and cutting way back on the tahini. A little hummus can be wonderful spread thinly on whole wheat toast and topped with tomato slices. It's also good served as a dip with whole wheat pita bread that has been cut into wedges and warmed.

SPEAKING OF THE MIDDLE EAST . . .

You'll find some of the same brand names here that you saw in the natural foods area of the store (in Aisle 1). For instance, you'll see more of the cup-of-soup-style meals, as well as a variety of mixes for such foods as tabbouleh salad, couscous, and pilafs. If you've never tried couscous, you're in for a treat. It's delicious, and it's low in fat. Couscous—tiny round pieces of semolina—is sometimes called "Moroccan pasta." It cooks quickly and its consistency is like grits or a coarse version of Cream of Wheat. A good way to serve couscous is as a main course mixed with chopped, steamed vegetables or pieces of chopped fruit.

With any of the mixes found in this section, cut way back on any oil that the directions say to add. Better yet, cut the fat out altogether. If you do, you may have to add a little more water, lemon juice, or other liquid to give the mix a bit more moisture.

ORIENT

In the Oriental specialty section, you'll find some versatile condiments. Tamari is a dark, rich soy

---

Talking shopping with Peter Burwash, tennis pro and Davis Cup winner, Honolulu, Hawaii:

If you met me at the checkout line in the grocery store, you might notice that the majority of my meals are based around ethnic-type preparations—whether Italian, Chinese, Thai, Indian, or Mexican. In the beginning, all societies were vegetarian. Only after they became affluent did they add meat to their diets.

In my travels to 134 countries around the world, I had the opportunity to sample various ethnic foods, and to be honest, I was amazed at the quality and flavor of the various ethnic-type foods. In the Western world, we've become dependent upon fats and sugars to make our food taste good. Because most of our food is aged and processed, we have to flavor it with fat, preserve it with salt, and add lots of sugar in order for people to enjoy the food. Overseas, however, herbs and spices flavor the fresh foods and provide an incredibly palatable meal.

sauce that is fat free (although high in sodium). Rice vinegar and mirin (sweetened sake) are used in some Oriental dishes. You'll find green tea, sweet-and-sour sauce, hoisin sauce (sweet and slightly spicy), and spicy brown bean sauce. All of these are low in fat and are fine to use as condiments.

You'll find a variety of stir-fry sauces here, too. Some contain oil, so read the labels and see where oil falls in the list of ingredients. Most of these are very low in fat. Even those that contain a tiny amount of fat can be fine to use when you add them to steamed fat-free vegetables and rice. When the sauce is used as a condiment in this way, the dish can still be very low in fat.

On the other hand, watch out for peanut sauce. This sauce is mostly fat, so avoid it.

You'll also find a variety of canned vegetables packed in water and salt—items such as bamboo shoots, baby corn, and water chestnuts. These are all great choices. Dried shiitake mushrooms and nori, a type of seaweed, are also sold here. My supermarket even sells a sushi-making kit. One piece of advice: stay away from sushi made with raw seafood—a great way to pick up a food-borne illness. If you want to try your hand at sushi, make it California style (veggie). Two other items that register high on the fat-o-meter are Japanese rice cracker mixes (those colorful mixtures of shiny, multishaped pieces) and fried chow mein noodles.

MEXICO

Did you ever think you would see the day when they would make fat-free tortilla chips? Well, that day has come. These chips are baked instead of fried. Give them a try with any of the salsas found in this section.

You'll find some varieties of canned beans in this section—an excellent choice that you'll see again on Aisle 3. Pear and guava nectar? Mix a little of these south-of-the-border juices with some seltzer water for a refreshing drink—an alternative to the same-old same-old.

You will also find flan mix here. Flan is a custardlike dessert traditional to many cultures; make it according to package directions using low-fat or nonfat milk instead of whole milk. I substitute low-fat soy milk for regular cow's milk, and it turns out wonderfully.

The Last Bite: Finding Exotic Ingredients and More International Foods

You may find, after experimenting with some of the ethnic food selections in the international section of your supermarket, that you enjoy them so much that you would like to work with ethnic cookbooks to create a wider variety of dishes. Your supermarket will carry some of the ingredients but not all of them. Where can you go to find a wider variety of exotic or international foods? In many communities with large populations of various ethnic groups, there are supermarkets that cater specifically to the needs of those people. For instance, in cities with large Indian populations, you may be able to find an Indian supermarket that sells traditional Indian foods and related items, such as many different types of lentils, spices, ready-made foods—even cookware. Check your local phone book to see what is available where you live, or ask the owner of your favorite ethnic restaurant. Interesting aromas and traditional music playing in the background can make some of these stores a unique experience.

GREAT CHOICES

- fat-free salad dressings
- fat-free mayonnaise
- flat bread
- most bread sticks
- ketchup
- low-fat marinades and barbecue sauces
- mustard
- pickles and pickle relish
- pickled vegetables
- vinegar—balsamic, herbed, fruited, and so on
- gourmet candies—fruit-flavored sours, hard candies, and pectin jelly beans
- imported fruit preserves and conserves
- German whole grain dark rye bread
- German red and white sauerkraut
- dried cherries and sun-dried tomatoes
- egg-free pastas
- artichoke hearts packed in water
- basmati rice
- chutney
- nan and chapati (Indian breads)
- grape leaves in brine
- *glykadi* (Greek vinegar)
- couscous
- sweet-and-sour sauce
- hoisin sauce
- spicy brown bean sauce
- many stir-fry sauces
- oil-free tortilla chips
- salsa
- pear and guava nectar
- flan mix (make it with low-fat or nonfat milk)
- canned beans—black beans, chickpeas, kidney beans

Talking shopping with Patti Breitman, literary agent, Fairfax, California:

Here's what you'd find in my grocery cart at the supermarket: vinegar (seasoned rice vinegar is my favorite salad dressing, and I use sweet-and-sour sauce for stuffed cabbage), lemons, carrots, celery, potatoes, lentils, brown rice, fruit (all varieties that look good), mixed salad greens, cabbage, whole wheat bread, a jar of roasted red peppers in water, garlic, and onions. With these staples—along with a jar of mustard—I can make endless varieties of sandwiches, stuffed cabbage, soups, salads, et cetera.

Aisle 3

Peanut Butter	Canned Vegetables
Rice	Canned Fruit
Soup	Gelatin

Is natural peanut butter really any better than regular style? Which soups are best? And how do all of those different kinds of rice compare nutritionally?

If you have ever had any doubt whether you could find oodles of good-tasting, low-fat foods at your supermarket, the last shred of concern should vanish once you look at all of the great choices available on Aisle 3. True, some of these foods do flash the yellow caution light, and there are one or two red-light foods here. But for the most part, there are green lights all the way down this aisle, with lots and lots of great choices available.

Peanut Butter and Jelly (and Other Spreadables)

PEANUT BUTTER

Okay, we'll tackle everyone's favorite first. Yes, peanut butter is mostly fat. As I mentioned earlier, if you use it, use it sparingly. Spread a *thin* layer on a sandwich, and add sliced bananas or apples with a sprinkling of cinnamon. Resist the temptation to add gobs of peanut butter to celery sticks and crackers (or to eat it by the spoonful from the jar).

Once we've said that, many people then wonder which is better—natural or regular-style peanut butter. In my opinion, it really doesn't matter. Use whichever you prefer—but again, use it sparingly. Why is the difference not important? Take a look at these two labels:

regular peanut butter

Ingredients: peanuts, sugar, partially hydrogenated vegetable oils, salt	Serving size: 2 T
	Calories: 190
	Fat (gm): 16
	Protein (gm): 9
	Carbohydrate (gm): 6
	Cholesterol (mg): 0
	Sodium (mg): 150
	Iron (% USRDA): 4
	Calcium (% USRDA): > 2

natural peanut butter*

Ingredients: peanuts, salt	Serving size: 2 T
	Calories: 200
	Fat (gm): 16
	Protein (gm): 8
	Carbohydrate (gm): 6
	Cholesterol (mg): 0
	Sodium (mg): 125
	Iron (% USRDA): 4
	Calcium (% USRDA): > 2

What's the difference between these two types of peanut butter? The regular variety has some sugar added. And the partially hydrogenated (that is, par-

---

*Some peanut butters—natural *and* regular styles—are available with no salt added. As I mentioned earlier, although salt or sodium intake is a concern for some people, most people can tolerate a moderate salt intake.

tially hardened) vegetable oil that has also been added helps to keep the peanut butter blended, so the oil doesn't rise to the top of the jar. (Natural peanut butter can be messy to deal with due to the half inch or so of oil that floats at the top of the jar.)

Nutritionally speaking, however, the two are actually about the same. Since peanuts themselves are mostly fat, the fat added to the regular peanut butter doesn't affect the total fat content of the product—it just displaces some of the fat that would otherwise be contributed by the peanuts. Tablespoon for tablespoon, natural and regular peanut butter each contain about the same number of grams of fat. The small amount of sugar added to regular peanut butter is also of little consequence.

Some people would argue that the partially hydrogenated oil added to the regular peanut butter is worse than the peanut oil by itself in the natural variety. In the context of the total diet, this point is not worth debating.

Again, if you use peanut butter only with the lightest hand, then the trade-off—the convenience of the regular peanut butter (it's less messy and you don't have to deal with a sea of oil at the top of the jar) versus the saturation level of the partially hydrogenated oil—is worth it. For practical purposes, these two jars of peanut butter are roughly equivalent nutritionally.

Incidentally, for anyone who has ever considered simply buying a jar of natural peanut butter and pouring off the oil that floats on top—I don't recommend it. I have tried it, and the result is a rock-hard chunk of peanut butter in a jar. You need a pickax to chisel it out.

There is a third option: reduced-fat peanut butter has recently become available.

reduced-fat peanut butter

Ingredients: peanuts, corn syrup solids, sugar, soy protein, salt, partially hydrogenated vegetable oils, minerals (magnesium oxide, zinc oxide, iron phosphate, and copper sulfate), vitamins (niacin, B_6, and folic acid)	Serving size: 2 T Calories: 190 Fat (gm): 12 Protein (gm): 8 Carbohydrate (gm): 15 Cholesterol (mg): 0 Sodium (mg): 140 Iron (% DV): 4 Calcium (% DV): > 2

Frankly, this product is not much different nutritionally from the other two. It contains about 25 percent less fat. That amounts to about four grams of fat per two-tablespoon serving. Extra sugar has been added, displacing some of the peanuts and resulting in less fat per serving. The cost of the reduced-fat peanut butter is about half a cent per ounce more, or ten cents on a one-pound jar. No matter which peanut butter you choose, the bottom line is still the same: use it very sparingly.

While we're on the subject of peanut butter, you may have seen the sweetened nut-butter spreads sold in many supermarkets. Some are chocolate-flavored. How do these compare to regular peanut butter?

Take a look at a label from a spread made with hazelnuts and cocoa, and compare it to regular peanut butter:

hazelnut and cocoa spread

Ingredients: sugar, vegetable oil, nuts, cocoa	Serving size: 1 T Calories: 80 Protein (gm): 1 Fat (gm): 5 Carbohydrate (gm): 9 Cholesterol (mg): 0 Sodium (mg): 10 Calcium (% USRDA): 2

regular peanut butter

Ingredients: peanuts, sugar, partially hydrogenated vegetable oils, salt	Serving size: 2 T Calories: 190 Protein (gm): 9 Fat (gm): 16 Carbohydrate (gm): 6 Cholesterol (mg): 0 Sodium (mg): 150 Calcium (% USRDA): > 2

The serving sizes listed for each item are different, so double all of the information given for the hazelnut and cocoa spread in order to make the serving size the same as it is for the peanut butter.

The hazelnut spread is just slightly lower in calories than the peanut butter. It's lower in fat, sodium, and protein, and it's higher in carbohydrates (because of the sugar added to it).

The most significant difference between the two products is probably the fat content. For a two-tablespoon serving, the peanut butter has six grams more fat than the hazelnut cocoa spread. The hazelnut cocoa spread is hardly low in fat, though. At ten grams of fat per serving, both products should be used very sparingly.

Talking shopping with Victoria Moran, author of *Get the Fat Out: 501 Simple Ways to Cut the Fat in Any Diet*, Kansas City, Missouri:

My advice for shoppers: for every item you pick up, ask yourself, "Is this a vegetable, fruit, grain, or legume?" If it's something else, be sure you plan to use it as a flavoring, garnish, or special-occasion food rather than as a major contributor to your diet. Also, reading labels is important, but the best foods around—fresh fruits and vegetables, dry beans, and basic grains like brown rice—don't even have them.

JAM, JELLY, AND SIMILAR SPREADABLES

I am most enthusiastic about jams, jellies, preserves, conserves—any of the variety of fruit spreads that you'll find on the supermarket shelves. These products are versatile, delicious, and fat free. Don't like seeds? Seedless jams are available. Prefer to avoid refined sugars? Fruit-only spreads are available with no sugar added. All of these products are similar in nutritional content, and assuming you don't eat them by the cupful (they don't provide much in the way of vitamins or minerals), they are all great choices.

Fruit jams, jellies, preserves, and the like can take the place of fatty spreads that you might otherwise add to bagels and toast. Start leaving the margarine off of toast, muffins, pancakes, and waffles, and load up with a heavenly fruit spread instead.* Keep a variety in your refrigerator. Supplement your standard grape-jelly diet with lime marmalade, fig preserves, or passion-fruit jam. Try red or green pepper jelly on crackers. Currant jelly and apple butter are other good toppings to add to your pantry. And there is a huge variety of beautifully packaged gourmet jams, jellies, and preserves available. Keep several on hand and ditch the cream cheese and butter.

More of the Sticky Stuff

MOLASSES, HONEY, AND SYRUPS

Aside from a hefty dose of iron in blackstrap molasses, these sugars are nutritional zeros. How-

---

*Recently, margarine has gained attention in the popular press due to research that suggests that "trans-fatty acids"—which are produced during the hydrogenation process that hardens vegetable oil into margarine—have an adverse effect on blood fat levels. More research is needed, but for now, more people are cutting back on margarine for this reason.

ever, like jams and jellies, they have their place. I am a big fan of a little sugar if it makes foods more palatable and takes the place of a fatty alternative.

Once again, leave the butter, margarine, and cream cheese off toast, muffins, pancakes, and waffles and use some honey or syrup instead. Most people prefer to use molasses in recipes, rather than eating it straight, but molasses can be used just like honey and syrup, if desired. All of these are great choices.

On the subject of syrups: did you know that most pancake syrups really contain only about 2 percent maple syrup? Corn syrup with coloring and a dash of maple syrup is a more accurate description. If you've never tasted pure maple syrup, give it a try. It is much more expensive than pancake syrup; to enjoy the taste but reduce the expense, try adding a little pure maple syrup to the supermarket pancake syrup for rich taste.

What about "light" syrups? Compare the following two labels:

regular pancake syrup

Ingredients: corn syrup, water, pure maple syrup (2 percent), artificial flavor, cellulose gum, sodium hexametaphosphate, sorbic acid and sodium benzoate (preservatives), caramel color	Serving size: 2 T Calories: 100 Fat (gm): 0 Protein (gm): 0 Carbohydrate (gm): 26 Cholesterol (mg): 0 Sodium (mg): 30 Iron (% USRDA): > 2 Calcium (% USRDA): > 2

light pancake syrup

Ingredients: sugar syrups, water, high-fructose corn syrup, cellulose gum, artificial flavor, salt, sorbic acid and sodium benzoate (preservatives), sodium hexametaphosphate, natural flavor, caramel color	Serving size: 2 T Calories: 50 Fat (gm): 0 Protein (gm): 0 Carbohydrate (gm): 13 Cholesterol (mg): 0 Sodium (mg): 90 Iron (% USRDA): > 2 Calcium (% USRDA): > 2

These two products are essentially the same nutritionally, with one difference. The light syrup has been diluted with water so that there is half the amount of sugar per serving. Because the calories in these pancake syrups come from sugar, the light syrup thus has half the calories.

Does this make the light syrup a better choice? Not necessarily. How much syrup would you use at a meal, and how often do you use it? Unless you use a lot of pancake syrup and you eat it often, the extra calories from the sugar are not likely to be a significant issue in the greater scheme of things.

Weight watchers may know that calories from carbohydrates, such as sugar, are not as much of a problem as calories that come from dietary fat (dietary fat is stored as body fat much more readily than are carbohydrates). Unless you are trying to control your weight and you use large amounts of pancake syrup, my advice is to use whichever syrup you prefer, for whatever reason.

Pancake and Waffle Mixes

Let's assume that you don't make pancakes from scratch. Which mix do you prefer? Here are three comparisons:

whole wheat pancake and waffle mix

Ingredients: whole wheat flour, enriched unbleached flour, oat flour, soy flour, brown sugar, leavening, dried molasses, wheat germ, dextrose, salt	Serving size: 3–4 pancakes
	Calories: 220
	Fat (gm): 8
	Protein (gm): 9
	Carbohydrate (gm): 30
	Dietary fiber (gm): 3
	Cholesterol (mg): 75
	Sodium (mg): 540
	Iron (% USRDA): 8
	Calcium (% USRDA): 20

The recipe on the box assumes that egg, 2% milk, and oil are added to the mix.

regular pancake and waffle mix

Ingredients: enriched unbleached flour, sugar, rice flour, leavening, salt	Serving size: 3–4 pancakes
	Calories: 180
	Fat (gm): 5
	Protein (gm): 6
	Carbohydrate (gm): 29
	Dietary fiber (gm): 1
	Cholesterol (mg): 70
	Sodium (mg): 550
	Iron (% USRDA): 8
	Calcium (% USRDA): 6

The recipe on the box assumes that egg, 2% milk, and oil are added to the mix.

light pancake and waffle mix

Ingredients: enriched unbleached flour, powdered cellulose, dried buttermilk, corn flour, soy protein concentrate with lecithin, sugar, leavening, dried egg whites, wheat gluten, salt, dicalcium phosphate, dried whole eggs, dextrose, partially hydrogenated soybean oil with mono- and diglycerides, nonfat dry milk, magnesium oxide, xanthan gum (and added vitamins and minerals)	Serving size: 3–4 pancakes
	Calories: 130
	Fat (gm): 2
	Protein (gm): 7
	Carbohydrate (gm): 25
	Dietary fiber (gm): 5
	Cholesterol (mg): 10
	Sodium (mg): 570
	Iron (% USRDA): 15
	Calcium (% USRDA): 35

The recipe on the box assumes that only water is added to the mix.

How would I call this one? The first step is to sift through the ingredients, taking into consideration any ingredients that would be added to the recipe. The instructions call for eggs, oil, and milk to be added to the first two mixes. The last mix already contains the equivalent of these ingredients in a lower-fat, lower-cholesterol form; only water is left to be added to the mix.

The last mix is higher in fiber than the others due to the powdered cellulose it contains. The first two mixes could be reduced in fat and cholesterol if the added egg and milk were replaced by an egg substitute and water or skim milk and if the oil were cut back by at least half. To replace the egg in the recipe, a commercial liquid egg substitute could be used, two egg whites could be used in place of one whole egg, or powdered egg replacer (from a natural foods store or the health food section of the regular supermarket) could be used. Or try the mashed banana trick described in The Produce Section.

All things considered, the best choice here is the first one: the whole wheat pancake and waffle mix, with the recipe altered by using an egg substitute, skim milk or water, and little or no added oil. Whole wheat flour is the first ingredient, and that is a better choice than the white flour and cellulose combination offered in the light mix (cellulose is of questionable nutritional value as a fiber source). However, any of these three mixes would be fine to use, providing the substitutions already mentioned were made in the first two mixes. Once you do that, then all of them are low in fat and cholesterol. Cook them on a nonstick skillet, and remember to skip the butter and to use syrup or a fruit spread on top instead. Warm fruit compote on top would also be delicious.

Rice

You'll find two types of products here: seasoned, packaged rice mixes and bags or boxes of plain rice.

Approach the seasoned and flavored rice mixtures carefully. Be aware of the following:

- Read the label. Many rice mixes may contain sulfites (preservatives) and/or MSG (monosodium glutamate), a flavor enhancer. Some people are sensitive to these additives.
- Many of these mixes call for margarine to be added. This can increase the fat content of the dish considerably. If you use these mixes, cut back on the fat or leave it out altogether. Moisture and flavor can be added by stirring in some vegetable broth or topping the rice mixture with sautéed vegetables.
- Nutrition information listed on the box may be "as packaged" or "as prepared." Be sure that you realize what you are looking at when you assess the nutritional composition of the dish.
- Some of the mixes are labeled "low salt," "low fat," and/or "low cholesterol." The recipes on these boxes call for less fat to be added than some of the other mixes. This may be the only difference between these and other mixes. When this is the case, you could just as easily modify one of the regular mixes to cut the fat content.
- Boxes labeled "one-third less fat" may still contain up to six grams of fat per half cup serving. That's a lot of fat, and a half cup is not a very big serving.

Some rice mixtures are flavored with seasonings that are unique to a particular ethnic cuisine, and these can be quick and convenient additions to a meal. Some bean and rice mixtures, for instance, are good—just leave out any added fat called for in the instructions. If the dish seems too dry as a result, you can consider mixing in a moist vegetable such as stewed tomatoes or sautéed greens.

A saffron rice mix is beautiful served with black beans that have been cooked with onions, garlic, and chopped red peppers. Curried rice mixtures add an Indian flair to a meal. Top the rice with curried sautéed vegetables. For a Spanish meal, try a paella rice mixture and steamed vegetables tossed together in a pretty casserole dish.

After all of this talk about seasoned rice mixes, there's still a lot to be said for plain old everyday rice. Rice is one of the most delicious, nutritious, and versatile foods. Plain rice of any variety is always a great choice.

There are many varieties of rice. What you'll see most often in the supermarket is long- or short-grained white rice, brown rice, wild rice (often mixed with regular long-grained rice), and basmati rice. As mentioned in Aisle 2 (page 104), basmati rice is an especially fragrant, flavorful variety of rice. Rice can take thirty minutes or longer to cook, but "quick" rices are also available that can take as little as five to ten minutes to cook. Quick rices have been pre-treated to make them cook faster. On the other hand, some people swear by their rice cookers—crock pot–like appliances that produce perfect rice and require no supervision!

Should you use brown rice instead of white rice? Generally, whole grains are preferable to refined grains. In processing refined grains, such as white

---

HELPFUL HINT

When you cook a pot of rice, make more than you need for that meal. Extra rice is great to use in low-fat rice pudding. Rice can keep in the refrigerator for at least a week and can easily be reheated in a microwave oven—add some to a plate of beans or cooked vegetables for a supernutritious, quick meal. Leftover rice can also be frozen until you want to use it.

rice, some of the fiber is removed, along with vitamins and minerals. White rice from the supermarket has usually been enriched—many of the nutrients removed during processing are added back. And surprisingly, there isn't all that much difference in the fiber content of brown rice as compared to white rice. Use whichever type of rice you prefer.

And Now . . . the Scoop on Soup

"Soup is good food." Well, maybe and maybe not. Most bouillons, soup bases, boxed soup mixes, and canned soups are exceedingly high in sodium, which may be a problem for some people. But many soups are relatively low in fat, even if they aren't nutritional powerhouses.

Read the labels for fat content, and be careful of some of the gourmet and cream soup varieties; these types most likely contain hefty amounts of fat and cholesterol. Some reduced-fat cream soups are now on the market, and these are better alternatives to the regular variety.

New on the market are some excellent cup-of-soup-style products, like those mentioned in Aisle 1. Some of these are made with lots of beans and lentils, and they're high in fiber and low in fat. Great choices.

One well-known brand of soup has a new line that is billed as being low in fat and cholesterol. Comparing the company's regular tomato soup to its "healthy" version of tomato soup, the ingredient lists are identical. In fact, the nutritional compositions are nearly identical as well. The only somewhat significant difference appears to be in the sodium content—460 milligrams of sodium in the healthy version and 670 milligrams in the standard soup. The healthy version costs about ten cents more per can. Either one of these products would be fine for most people.

Mixed in with the soups in this section are a variety of sauce mixes for such dishes as goulash, sauerbraten, and coq au vin. Most of these are designed to be used with fat-laden, meat-based dishes. But if you are considering buying one of these mixes to use in some other creative way, plan to modify the recipe to reduce any added fats, such as cream, oil, butter, or whole milk.

Another product that you may find in this section is ramen noodles. Ramen noodles are Oriental-style noodles that are frequently sold dried and wrapped in a bundle (resembling a bird's nest); they are packaged in a plastic wrapper and sold in the Oriental foods section of the supermarket or on the soup aisle. They are usually seasoned with spices and/or pieces of dried meat, poultry, or seafood. When boiling water is added, the ramen noodles become a sort of noodle soup.

Ramen noodles are frequently fried before being packaged, so some can contain a fair amount of fat. Some may also contain monosodium glutamate (MSG), which some people may wish to avoid. So read labels carefully. Others are lower in fat and may only contain a few grams of fat per serving. The natural foods section of the supermarket may also sell ramen noodles to which no meat or MSG has been added. Ramen noodles are a good choice if they are reasonably low in fat or fat free.

Gelatin, Pudding Mixes, and Snack Packs

Let's get this one out in the open: gelatin is *not* food. If ever there was a nonfood food, gelatin is it. Primarily a vehicle for whipped cream, sour cream, and the occasional fruit cocktail, gelatin mixes are essentially a fat-free blend of sugar, gelatin, artificial flavor, and artificial color. Some are even sugar free,

---

Talking shopping with Debra Wasserman, codirector of the Vegetarian Resource Group, Baltimore, Maryland:

From Memorial Day to Labor Day, I tend to buy a lot of fresh produce at the local farmer's market; I can get organic items at a good price. During the remainder of the year, I buy the following at the supermarket:

fresh fruits and vegetables; some frozen vegetables
fruit juices
pasta and plain tomato sauce
various breads
canned beans (garbanzo, kidney . . .)
applesauce and fruit jams
spices and flour
cereals

---

sweetened with artificial sweeteners, thus removing the only nutritive ingredient that was ever in the stuff.

Gelatin mixes are fat and cholesterol free; they have that much going for them. However, much like soft drinks, they have no nutritional merit. They're chemical soups, and I avoid them.

Similar to gelatin mixes, pudding mixes are fat free unless low-fat or whole milk is used in preparing them. In contrast to gelatin, they do have some nutritional value since they are made with milk. However, most pudding mixes contain a slew of artificial flavors and colors. The no-bake pie fillings and cheesecakes found in this section are similar, but the end products are usually high in fat.

Snack packs—individual-sized containers of gelatin and pudding—may be low in fat or even fat free, but these, too, are loaded with artificial flavors and colors. On balance, I generally do not recommend these products. They're fiberless and full of chemical additives of questionable safety. These products will also usually constitute a significant portion of someone's daily intake—people eat a bowlful—in contrast to a few hard candies or a tablespoon of jelly that, while not nutritious, tend to be eaten in small quantities.

One very nice alternative to pudding mixes is tapioca. Simple to prepare, tapioca pudding can be made using skim milk or low-fat soy milk. Added bits of dried fruit can be delicious, too.

And Speaking of Dried Fruit . . .

I promised you loads of great choices on this aisle, and here are a bunch. Dried fruits are delicious and nutritious, and they can make an ordinary dish much more interesting. One caveat: brush your teeth after eating dried fruit, because it sticks to the teeth and may promote cavities. And many packages contain sulfites, preservatives to which some folks are sensitive (shop for sulfite-free dried fruits if you are sensitive). Generally speaking, though, these are a great choice.

You don't have to be constipated to enjoy prunes. Some are flavored with natural lemon or orange. They're high in iron and full of fiber. Chopped dates are absolutely wonderful cooked with hot oatmeal—add a dash of cinnamon, too. Dried apples, apricots, figs, currants, raisins, and mixed dried fruit bits can all be added to hot or cold cereals, muffins, and breads. Be creative. Add dried fruit to salads and casseroles. A handful of raisins tastes great in vegetarian chili.

Fruit in Other Forms

If you like canned fruits, go ahead and indulge. True, fresh fruit has more fiber (mostly because it's likely to have a peel, like peaches or pears). And canned fruit does lose some of its nutritional value in the canning process. But all in all, canned fruits are usually nutrient dense while being fat and cholesterol free. They are a great choice.

Mandarin orange segments or grapefruit sections look pretty and taste good in green salads. You will see canned peaches, pears, apricots, fruit cocktail, and pineapple here, and you'll usually have a choice of fruit packed in heavy syrup, light syrup, or no syrup. Fruit packed in its own juice is the best, but all are acceptable, unless you have a medical reason for avoiding added sugars.

Another great choice is applesauce, and there are some nice new varieties, such as those made with McIntosh and Granny Smith apples. Some applesauce is mixed with other fruit, such as strawberries or cherries. Some has cinnamon added. Canned fruit snack packs are a much better choice for bag lunches and snacks than the gelatin and pudding options described earlier.

You'll also see some canned-fruit pie fillings in this section of the store. If you do use these, look for fillings that are simply fruit, water, and sugar, without artificial flavors and colors added. Canned pumpkin is a terrific choice—it's just full of vitamins and minerals. Canned cream pie fillings are probably high in fat. There may not be any nutrition information on the label, but notice that oil

---

HELPFUL HINT

Do you love pumpkin pie? Try making the filling by itself. Cook it in a shallow baking pan and forget about the crust (which is usually high in fat)—sort of a pumpkin pudding instead. If the recipe calls for milk and eggs, substitute skim milk or low-fat soy milk, and use an egg substitute (see Aisle 11).

or other fats are usually added to these.

Items such as spiced crab apples or apple rings, and maraschino cherries are to be used as garnishes or condiments. Maraschino cherries typically have artificial color and flavor added.

Potatoes—the Quick Kind

Some people are devoted to instant mashed potatoes. They're fine, unless you are sensitive to sulfites, which most brands contain. Other potato mixes, however, can call for a lot of added fat. Read labels carefully on such products as au gratin potatoes and scalloped potatoes. Actually, fresh potatoes can be just as quick to prepare as some of these mixes. See The Produce Section (page 60) and Aisle 2 (page 99).

Beans—Dried, Canned, and Mixes

I was raised in Michigan, where my exposure to beans was fairly limited. It wasn't until I moved to the South that I really became hooked on them. A huge variety of beans is available. They are beautiful, delicious, and versatile. You may find that specific types of beans may be more readily available in some parts of the country than in others. But here's a partial list of the beans, dried and canned, that I found on my southern supermarket shelves:

  pinto beans
  lima beans
  black beans
  red kidney beans
  split green peas
  Great Northern beans

garbanzo beans
small white beans
navy beans
black-eyed peas
speckled lima beans
lentils
adzuki beans
vegetarian beans in tomato sauce
pork 'n' beans
white kidney beans
field peas with snaps
chili hot beans
plus many seasoned bean mixes

Beans are loaded with fiber—about sixteen grams in a cup. Current dietary recommendations call for at least twenty-five to thirty-five grams of fiber in your diet every day (most Americans only get about ten grams per day). Beans can bring you a long way toward meeting fiber goals. Plus, they're full of vitamins and minerals and are low in fat. They are a fantastic choice. Some beans, such as traditional baked beans, are made with pork or other meat. Try to buy meatless varieties. However, even the beans that include some meat flavoring are still low in fat overall.

Bean mixes are available that combine one or more types of dried beans with seasonings—herbs and spices—to create a special bean dish. You'll also see dried, multibean soup mixes in the supermarket. These bean mixes all tend to have salt added, so be aware of that if you are trying to limit your sodium intake. Otherwise, these mixes can be wonderfully delicious and nutritious.

As for the case for canned beans versus dried: it really boils down to a matter of personal preference. Dried beans have to be soaked—usually overnight—before they can be cooked. An exception is lentils or split peas, which are tiny and have lots of surface area, so they cook relatively quickly, usually without soaking. If you don't like to plan ahead or simply can't be bothered, then canned beans may be a convenient alternative to dried beans.

True, canned beans are usually packed with salt. But once again, if sodium is not a major issue for you, then it may not matter. On the other hand, if you need to reduce your sodium intake but still want the convenience of canned beans, then rinse the beans in a colander before you cook them. Rinsing them will wash away most of the added salt.

As for the vitamin and mineral content of canned beans as compared to dried: it's true that for some foods, canning destroys certain nutrients (such as vitamin C). Vitamins and minerals can also leach out of the canned vegetables over time and end up in the water in which the vegetables are packed. That's the reason some people like to save the water from canned vegetables and use it in recipes or as soup stock. The nutrient loss from beans in the canning process is not as great as for some other vegetables, though. Canned beans are still a super source of nutrition. Go ahead and use them if you enjoy the convenience.

HELPFUL HINT

Add a can of beans to your favorite soup. For instance, if you heat up a can of tomato soup, dump in a can of navy beans to add tons of fiber and great taste. Here's another idea: take a basic dry soup mix, such as an envelope of tomato-basil soup. Begin heating it in a pan over the stove, and while it's heating, add two or three different fifteen-ounce cans of beans, such as garbanzo, pinto, or kidney beans. You can even toss in leftover chopped vegetables, or add a big handful of frozen mixed vegetables.

Other Canned Vegetables

Maybe canned vegetables sound a little mundane, but they are all great choices. True, they aren't quite as nutritious as fresh or frozen vegetables. Some vitamins and minerals are lost due to heat in the canning process, and some leach out into the water in the can. Nevertheless, some people may prefer canned vegetables to fresh or frozen, and they are acceptable. Despite some loss in nutritional value, they are still good fiber sources, they're usually fat free, and they still contain plenty of vitamins and minerals.

Canned vegetables usually have salt added, although there are also lots of salt-free varieties available now. Just like canned beans, canned vegetables can be rinsed to remove any added salt. One more note: overcooking *any* vegetable destroys vitamins. Canned vegetables should be the last item to be cooked before a meal is served, so that they don't sit on the stove turning to mush while the rest of the meal is still being prepared. They should be heated just until they are hot enough to eat—don't overcook. Microwaving can be a quick way of heating vegetables and preventing some nutrient loss.

HELPFUL HINT

Try eating canned greens—collards, turnips, spinach, mustard greens, or mixed greens—southern style: heat them and serve them with a little vinegar drizzled on top. Add a dash of hot sauce, too!

GREAT CHOICES

- fruit preserves and conserves
- fruit-only and low-sugar spreads
- jam
- jelly
- syrup
- honey
- molasses
- whole wheat pancake and waffle mix (with recipe modified for eggs, milk, and oil)
- light or regular pancake and waffle mixes (with recipe modified as noted)
- all varieties of rice (without added fat)
- low-fat bean and lentil soups
- most canned soups or soup mixes, except cream soups (some low-fat cream soups are now available)
- tapioca pudding made with skim milk or low-fat soy milk
- dried fruits—prunes, apples, apricots, raisins, currants, figs, dates, and mixed fruit bits
- canned fruits—peaches, pears, mandarin oranges, grapefruit sections, cranberry sauce, fruit cocktail, apricots, pineapple
- applesauce
- canned-fruit pie fillings
- canned pumpkin
- instant mashed potatoes (but may contain sulfites)
- all types of beans and peas (dried or canned)—pinto beans, black beans, red kidney beans, split green peas, garbanzo beans, navy beans, lentils, and so on
- all types of canned vegetables (no oil added)

Talking shopping with Deborah Madison, author of *The Greens* and *Savory Way* cookbooks, Santa Fe, New Mexico:

I am still a perimeter shopper at my supermarket, even though I have tried hard to use the inside aisles more. I go in the door and go immediately to the produce. And that's mainly what I buy. Actually, I prefer to shop for organically grown vegetables at another store, but since I'm often testing recipes and need things that my regular store may not carry, I go to the super. I *love* broccoli rabe, and I buy it virtually every time I shop. I could eat it every day.

As for the inside aisles, I've tried but don't find much. What I tend to buy are the same things I'd get in a health food store—good-quality grain cereals—oatmeal, cornmeal, grits, rice cakes, crackers. I occasionally get canned tomatoes and use those for pasta sauces. I keep a can of chickpeas on hand but find that other beans, including the dry ones, aren't so great—though living in New Mexico, we can get good pinto beans in bulk.

Aisle 4

Canned Meats	Coffee
Spaghetti Sauce	Kosher Foods
Pasta	Mexican Foods

This is the aisle where you'll find the pasta toppers, macaroni-and-cheese mixes, and hamburger helpers. It's home to the kosher products, coffee, tea, cocoa, and much more.

Canned Encounters
of the Mysterious Kind

May I be blunt? There are products on this aisle that I've heard about and never tried—canned meats so greasy and fatty that I don't know how they could be appealing to anyone, particularly in canned form. And nutritionally, they're real losers. But apparently they are quite popular, especially in the region where I live, which I understand is the Spam-eating capital of the world. Here, canned meats make up a fairly good-sized section of this aisle. I'm talking about such items as canned sausages and the aforementioned delicacy.

Amid the many ultra-high-fat choices, there are a few that are low in fat and cholesterol. Water-packed tuna is one example. There are also a few other kinds of water-packed seafood. The bottom line here is what we have said before: if you include some low-fat meats in your diet, the key is to use them as condiments. Refer back to The Meat Counter for an explanation of how meats and other animal products fit into a health-supporting diet. For the purposes of this book and for anyone striving for a low-fat, plant-based diet, I recommend avoiding these products completely.

Mixed in with the canned meats, though, is a product that might be convenient if used creatively. Take a look at this:

all-natural sloppy joe sauce

Ingredients: tomato concentrate from red, vine-ripened tomatoes, sugar, vinegar, salt, dehydrated onions, dehydrated peppers, spices, guar, xanthan, and carob bean gums, natural flavorings	Serving size: ¼ C Calories: 30 Protein (gm): > 1 Fat (gm): 0 Carbohydrate (gm): 6 Cholesterol (mg): 0 Sodium (mg): 360 Calcium (% DV): > 2 Iron (% DV): 4

If you like the flavor of sloppy joe sauce, you might use this to create your own healthy version of the old favorite. Instead of using lean ground beef or ground turkey to make the sloppy joes, why not try mixing the sauce with chopped vegetables and beans? Add some bulgur wheat to give the sauce the texture of ground meat. Leftover rice with chopped vegetables might also be good. Use your imagination. Remember, you are creating new traditions when you choose to eat more healthfully. New dishes don't have to taste exactly like the ones they are replacing; they just have to taste *good*.

(A note just for fun: Spam now comes in a "light" version, with "25 percent less fat." How much fat is that? Would you believe *twelve grams* of fat in one skinny little two-ounce serving?!)

Canned Meals and Lunch-Box Take-Alongs

You may remember the canned spaghettios—
spaghetti and tomato sauce—from when you were a
kid. But if you don't have kids of your own, you
may not have noticed how far spaghettios have
evolved. Similar to cup-of-soup-style lunch-box
portables, there are now single-serving containers of
pasta-and-tomato-sauce meals that are ready to be
microwaved at work or at school. Some are particularly kid-friendly, with the pasta cut in cute shapes
and packaged in wildly decorated containers.

Nutritionally, these rank from not-so-good to fair,
depending on your criteria. One example contains
three grams of fat for a seven-and-a-half-ounce
serving. Not too bad. But then there's the chicken
fat and the added MSG on the ingredient list. No
thanks. Another cheesy macaroni-and-cheese kid's
meal contains eleven and a half grams of fat in the
single-cup serving. Wow. No fiber here, either.

Only one item—a cheese ravioli in tomato
sauce—contained only one gram of fat in the single-serving cup and contained no other objectionable ingredients. For little kids and big kids alike,
the nutritious, fiber-rich bean and pea soups, chili,
and bean and rice mixtures that are available now
(and discussed in Aisle 1 and Aisle 3) in single-serving, microwavable cups are a better choice for a
quick and portable meal.

Seasoned Pasta Mixes, Macaroni-and-Cheese Dinners, and Hamburger Helpers

Like the rice mixes described in Aisle 3, seasoned
pasta or noodle mixes and macaroni-and-cheese
dinners are typically prepared by adding margarine

and/or milk, which boosts the fat content of these dishes. A Parmesan cheese angel-hair pasta mix calls for one tablespoon of margarine or butter to be added to the mix. "As packaged," this mix contains three grams of fat. But after the margarine is added, the dish totals six grams of fat in a half-cup serving. Another noodles-and-Alfredo-sauce mix totals ten grams of fat in a half-cup serving, or seven grams if the dish is prepared with skim milk and only half of the margarine.

The primary problem with most of these pasta mixes is the amount of fat they contain. Note that the nutrition information given is for a half-cup serving. Is a half cup a realistic serving size? If the dish is being served as an entrée, then my bet is that the serving size is really going to end up closer to one cup or more—so the fat is going to be double the amount on the label, or more.

This could easily mean a fat intake of at least twelve to twenty grams, just for that portion of the meal. That's a lot of fat for most people. A little further on in this chapter, we'll look at easy, convenient, low-fat ways to season plain pastas, as an alternative to these fattier pasta mixes.

Pasta salad mixes are also on the market. Comprised of dried pasta, sometimes some dehydrated vegetables, and a packet of salad dressing, the salad dressing can make or break the dish in terms of the fat content. One variety contains seven grams of fat in a serving, which is listed as "one-sixth of the box." It is hard to judge just how much—and how realistic—that serving size is without actually preparing the mix once.

A light version is also available. The main difference here is that a reduced-fat dressing is used, so the fat content falls to only one gram per serving. This fat-reduced pasta salad can be an okay choice, if you don't mind the artificial colors and assorted preservatives that happen to be included in the mix.

Our last example, a macaroni-and-cheese mix, is almost humorous: one very popular brand lists the fat content of the dish as being two and a half grams, with a small asterisk following. The devoted label reader then has to look at some small print in the next column to read that the "2.5 grams" refers to the amount of fat *in the box only*. When prepared as instructed, one serving (one cup) contains "an additional" 15.5 grams of fat. Why not just say it up front—"one serving contains 18 grams of fat"—and be done with it?

What about meat extenders, such as the hamburger helper variety? Using grain or rice mixtures to make meats go a little further is a step in the right direction. However, as in the example given of the sloppy joe recipe, the best choice is to move away from dishes that call for meat as a main ingredient and to replace these with low-fat, meatless alternatives. The cookbooks listed at the back of this book are a good source of ideas. Another option might be to use these products with the soy- or vegetable-based meat-substitute products now available, particularly those that are fat free; these are discussed in The Meat Counter.

Pizza Mixes and Boxed Italian Dinners

Boxed pizza mixes are made up of three parts: the pizza dough mix, a packet of grated cheeses, and a container of pizza sauce—sort of a paint-by-number approach to making a pizza, with ingredients that could easily be purchased separately and assembled just as quickly. There are ready-made pizza crusts in the refrigerator section of the supermarket (see Aisle 12), and you can also buy the boxed

Talking shopping with Mollie Katzen, author/illustrator of *Moosewood Cookbook*, *Enchanted Broccoli Forest*, *Still Life with Menus*, **and** *Pretend Soup*, **Kensington, California:**

I shop in three or four different places for my groceries. Here in the San Francisco area we have some big produce stores where you can buy all sorts of organically grown, fresh produce and not spend more than twenty dollars. I do go to the regular supermarket for my children (ages three and nine). There I typically buy:

little boxes of Motts apple juice
applesauce
lots of bottled fruit juices
nonsugary breakfast cereals
some hot cereals
some health food–brand cereals
bagels that are delivered fresh from a bakery
granola oat bread (my kids love it)
peanut butter (health food brand)
Boboli pizza forms
frozen yogurt or frozen fruit bars for my kids
occasionally, canned beans (health food brand)
tea
Edensoy soy milk for my daughter
lemonade
mineral water
organic tofu
organic fruits and vegetables if available
baking supplies

pizza crust mix by itself, without the cheese and sauce.

Probably the best quick-and-convenient way to make your own pizza is to buy a ready-made crust (preferably whole wheat) or make one from a mix, then add your own toppings, going heavy on the chopped fresh veggies and tomato sauce and leaving off the cheese. Like I said before, if a cheeseless pizza sounds really radical, keep an open mind and give it a shot. I was genuinely surprised when I tried my first cheeseless pizza—it was delicious. (See also The Deli/Bakery earlier in the book.)

Boxed spaghetti and lasagna dinner "kits" are similar. If you don't mind buying the components of these dishes individually, you can have a lot more control over the nutritional content of the meal. For instance, you could buy whole wheat pasta instead of pasta made from refined white flour. You could also replace full-fat cheeses with nonfat varieties, or you might skip the cheese altogether.

One popular spaghetti dinner mix contains three grams of fat in a five-ounce serving, with four servings per carton. Think about how many servings you would actually consume at one sitting when you evaluate the fat content. It may be hard to visualize until you actually prepare the recipe once and see how much it makes. The label on a lasagna dinner mix says it contains eight grams of fat in one six-ounce serving, and the container makes four servings. These dinners may be okay on occasion, depending on how much you eat and what else you serve with the meal.

Bottled Sauces and Dried Sauce Mixes

You'll see a variety of bottled sauces at the supermarket—sauces meant to be served on top of beef and chicken dishes, numerous tomato-sauce blends for pasta and other entrées, and the odd cheese sauce or gravy. Nutritionally, these can range from great to pretty dismal. Let's focus on the great, starting with some bottled spaghetti sauces and less-fancy canned tomato sauces.

BOTTLED AND CANNED TOMATO SAUCES

Who hasn't used a bottled spaghetti sauce lately? Judging by the amount of shelf space devoted to these time-savers, I'd say just about all of us use them. There are so many varieties to choose from, too—how does spicy red pepper tomato sauce sound? Or onion and garlic, mushrooms and olives, or tomato basil? They're not just for spaghetti, either. These bottled sauces can be used on pizza, in numerous pasta dishes, on casseroles—you name it.

Okay, take a look at the label information for one typical bottled spaghetti sauce:

typical mushroom spaghetti sauce

Ingredients: tomato puree, diced tomatoes, corn syrup, vegetable oil (one or more of the following: corn, cottonseed, or canola), mushrooms, salt, spices, onions, garlic, citric acid, parsley, spice extract	Serving size: 4 oz.
	Calories: 120
	Protein (gm): 2
	Fat (gm): 4
	Carbohydrate (gm): 20
	Cholesterol (mg): 0
	Calcium (% USRDA): 4
	Iron (% USRDA): 4

The ingredients in this sauce are fairly wholesome, but enough vegetable oil has been added that one serving of this sauce contains four grams of fat. That's not an alarming amount, but there are some sauces available now that are fat free. Look for them. Also note that some bottled spaghetti sauces contain meat. Buy the vegetable-flavored sauces or marinara sauce, instead.

Personally, I find many of the bottled spaghetti sauces to be too rich and sweet tasting when I use them straight from the bottle. So I dilute the bottled sauce with some plain canned tomato sauce. This can also cut down on the amount of fat in a serving substantially. See the Helpful Hint that follows.

> **HELPFUL HINT**
>
> Dilute bottled spaghetti sauces with plain canned tomato sauce to balance out a too-rich or too-sweet flavor and to reduce the fat content of the sauce. For example, mix your favorite bottled spaghetti sauce one-to-one with plain canned tomato sauce. An alternative method is to pour a can of plain tomato sauce into a mixing bowl, then add the bottled sauce one scoop at a time, stirring and tasting after each addition, until the mixture tastes right to you.

Some canned tomato sauces (and canned crushed tomatoes) are also seasoned with Mexican or Italian spices. I spotted one Mexican-style sauce with no fat at all. An Italian-style tomato sauce had a little olive oil added, so it contained one gram of fat in a quarter of a cup. Not bad. Tomato purees and tomato paste are also fine.

THE REST OF THE PACK

Aside from the multitude of spaghetti sauces out there, a fair amount of supermarket shelf space is devoted to a collection of other sauces. Some are bottled, such as a creamy whitish chicken cacciatore sauce that contains two grams of fat in a half cup. At my store, I saw a sweet-and-sour sauce that is fat free, and a light honey mustard sauce with less than one gram of fat per half-cup serving. The labels on these sauces indicate that they are to be used in meat dishes.

In all of these cases, it's not so much the sauce that is the problem as the concept of putting the sauce over some type of meat. Here's where that mind-set comes in again: the challenge is to move away from meat-centered meals and develop new eating habits that center on meals made from whole grains, vegetables, legumes, and fruits. And just as in the example of the sloppy joe sauce given earlier, many supermarket items can be used in nontraditional ways. Use your imagination and experiment a little.

The same goes for the huge array of dry mixes that you'll find in this section. There are mixes for gravies, stir-fry seasonings, rice seasonings, cheese sauce, white sauce, chili seasoning, taco seasoning, and so on. Some of these mixes contain ingredients such as sulfites and MSG, to which some people are sensitive. Others call for milk and/or margarine, butter, or oil to be added (use skim milk, low-fat soy milk, or water instead and cut back on or eliminate the margarine, butter, or oil). Some packages contain nutrition information; others do not. If you decide to use one of these mixes for convenience's sake, keep all of these points in mind, and let your imagination guide you in using some of them in nontraditional ways.

Pasta, Pasta, and More Pasta

Rainbow rotini, vermicelli, mostaccioli, ziti, orzo—do any of these spell "pasta" to you? How about lasagna, elbow macaroni, linguine, fettuccine, angel hair, rigatoni, or seashells? There are an incredible number of types of pastas and an equally large number of delicious ways to fix them.

We started to discuss pasta earlier in this book (see The Deli/Bakery). Some supermarkets sell fresh pasta in the deli area. Then there is usually a large section of dry pastas elsewhere in the store (Aisle 4, in this case) and a smaller section of refrigerated fresh pasta from which to choose.

In the deli area, I warned against buying fresh pasta made with egg yolks, because of the high cholesterol content. The best type of pasta to buy is made without eggs. However, there are also some varieties of fresh pasta—found in the refrigerated section of this aisle or elsewhere in your store—that are made with egg whites (not whole eggs), so the cholesterol content is not a problem. You may also see fresh cheese ravioli and tortellini. These tend to be higher in fat due to the cheese used; they are not a great choice.

Don't think that good pasta choices are limited, though. There are a growing number of egg-free pastas available now. Check the labels. Many are gourmet style. How does tarragon and chives, wild mushroom, tomato basil, or spinach nutmeg pasta sound? My supermarket also carries egg-free saffron pasta, lemon and pepper, garlic and parsley, Cajun, and confetti pastas. Believe it or not, it even sells *chocolate* pasta (what type of sauce would you serve with *that*?).

You might also see boxes of dried mini ravioli and tortellini. These products contain dried cheeses that are high in fat. According to the package, a two-ounce dry serving of either of these contains six grams of fat. There are four servings in a box.

Your next question might be, How big is a serving? The box doesn't say. You would have to fix this dish once in order to determine how big a serving one-fourth of a box comes out to be. Only then could you judge how realistic the serving size is and how much you would actually eat at one meal. Six grams of fat is probably an acceptable amount for most people if the other components of the meal are low in fat or fat free.

What about pasta toppers? We covered the subject of tomato sauces a little earlier. How about some others?

Pesto sauce is popular. It's usually made with olive oil, basil, and Parmesan and/or Romano cheese. It may be sold in bottles, refrigerated fresh, or in a dry mix. Pesto is a rich, flavorful, aromatic seasoning for pasta, but it's also very high in fat. The fat content can be controlled somewhat by using a dry mix and substituting water for part or all of the olive oil called for in the directions. You may find the result tastes rather flat, though, and you might be better off simply tossing the pasta with a little Parmesan cheese and chopped, fresh basil leaves instead.

Isn't Parmesan cheese high in fat? you might ask. The answer is yes, but used as a condiment—a sprinkling over a plate of pasta—the fat content doesn't add up to much. Some people like to use nutritional yeast in place of Parmesan cheese. It has a pungently flavorful taste. Nutritional yeast can be found in natural foods stores.

If you must use regular pesto, then bottled or fresh pesto should be used very sparingly, tossing a plate of pasta with only a teaspoon or two of pesto sauce. Since it's so flavorful, a little bit of pesto can really go a long way. I spotted a new variety of pesto sauce in my supermarket, too. This one was made with sun-dried tomatoes instead of basil. Use it the same way as the basil pesto if you want to give it a try.

Others? Well, Alfredo sauce is generally very high in fat. Check out the Helpful Hint that follows for a few ideas.

There's one last trick that can help if you absolutely must indulge in a dish topped with Alfredo or pesto sauce. It's the old salad-dressing trick applied to pasta: instead of ordering pesto pasta or fettuccine Alfredo the usual way, ask for the sauce on the side. Then, take a forkful of pasta and lightly dip it into the dish of sauce. All you need to do is *touch* the pasta to the sauce. You'll get the flavor of the sauce with a fraction of the fat. Heck, if you can't decide between pesto and Alfredo, you can always order a small dish of each. Some people even like

---

HELPFUL HINT

Looking for some good pasta toppers? Try these:

- Top cooked pasta with a medley of steamed fresh vegetables. Add a sprinkling of Parmesan cheese if you want to.
- Spaghetti sauces seasoned with herbs and vegetable flavorings are good. Dilute commercial, ready-made sauces with plain tomato sauce if you prefer. Add some chopped, cooked zucchini, yellow squash, broccoli, and carrots.
- Toss pasta with diced, fresh Roma tomatoes (remove the seeds first), chopped fresh basil, and a sprinkling of Parmesan cheese. An original recipe called for chopped walnuts and shredded mozzarella cheese to be added to this dish, which add a great deal of fat. An alternative would be to replace these with a little grated, fat-free cheese and/or some crunchy, chopped water chestnuts or other finely chopped vegetables.

---

to mix marinara sauce with Alfredo sauce to make a "pink" sauce. And at the end of your meal, you'll be amazed at how little sauce you have actually eaten—you won't even have made a dent in the bowlful. And to think that all of that sauce might have been on your plate . . .

Kosher Foods

Kosher foods are foods that have been sanctioned according to Jewish law. Most supermarkets have a section devoted to these products, but you don't have to be Jewish to enjoy some of them. Many of these items have counterparts in various other sections of the supermarket—Oriental food items,

Talking shopping with Cassandra Peterson, actress who portrays the character Elvira, Mistress of the Dark, Hollywood, California:

Pasta is one thing that I do buy at the regular supermarket, because I'm turned off by the whole wheat pasta sold at natural foods stores. It has a weird texture. I lived in Italy for a year and a half and learned to cook there. I love to eat pasta with a variety of different sauces that I make myself. I usually make a tomato-based sauce, spiced up with things like hot, dried chili peppers or capers. I also make a pureed broccoli sauce and add arugula, which is a green sort of like a dandelion green. It's tart and peppery-tasting. So, my pasta sauces are vegetable-based. I never use cream or butter, but I do use some olive oil and garlic.

soups, bread crumbs, egg noodles. However, you'll find a few items worth noting here that you won't find anywhere else in the store.

One of these items is the matzo. Matzo is unleavened bread that is traditionally eaten by Jews during Passover. Sometimes it is ground finely into matzo meal or matzo flour and used to make other foods, such as matzo balls. Matzos look like very large, thin crackers and are packaged in big squares wrapped in paper and plastic. They are available made from whole wheat flour or refined white flour. The ingredient list is very short—flour and water. Whole wheat matzos are made with stone-ground whole wheat flour, bran, and water. Some are salted, and some are unsalted. Matzos are a great choice.

Another great choice that may be new to you is borscht, which is sold bottled in this section of the store. Borscht is a Russian beet soup made with water, beets, sugar, and citric acid. It can be served chilled or heated. Just avoid the dollop of sour cream that is traditionally served with this soup. It's great all by itself.

The potato pancake mix in this section is made with sulfites to keep the potatoes white, and the instructions call for frying the pancakes in oil. There is no nutrition information available, but the list of ingredients includes potatoes, potato starch, salt, onion, and vegetable shortening (partially hydrogenated cottonseed oil). Fat is the last ingredient, but it's still tough to figure out how much of it the mix really contains and if it's a significant amount. See Aisle 3 for help with pancake mixes. If you use this mix, fry the pancakes in a nonstick skillet, rather than in oil. When you suspect that an item may be high in fat but can't be sure because of lack of label information, I recommend avoiding the product, as in this case.

You'll find an item called halvah here as well. (I wish I had a nickel for every time someone said my name reminds them of "that dessert called halvah.") Halvah is a delicious but extremely high-fat sweet made with crushed sesame seeds, sugar, oil, and other ingredients. A better treat from this section would be one of the kosher fruit fillings or toppings that can be found here, such as prune, pineapple, red raspberry, cherry, or apricot. A scoop of one of these on top of a matzo would be a great choice if you need something sweet.

HELPFUL HINT

Some of you may be interested in *The Low-fat Jewish Vegetarian Cookbook: Healthy Traditions from Around the World*, by Debra Wasserman (Baltimore, Md.: Vegetarian Resource Group, 1994). It's available at any bookstore. If you don't see it on the shelf, ask the store to order it for you, or order it directly from the Vegetarian Resource Group (the address is listed at the back of this book).

More International Foods

We found international foods on Aisle 2, but there are some additional items here. Some good Oriental food choices include bean thread noodles, which are made with mung-bean starch and water, and fortune cookies, which are low in fat.

Some additional Mexican food items can be found here, too. Canned refried beans are delicious, nutritious, and very versatile. Contrary to the way it sounds, refried beans are usually low in fat, even when some fat has been added to the can. Regardless, you should still seek out the fat-free varieties that are now available. Old El Paso is one example of a brand marketed nationally that now offers a fat-free version. Avoid those that contain lard. At this time, most canned refried bean products do not give nutrition information on their labels.

Refried beans can be heated and used to make bean burritos, bean tacos, bean sandwiches, and bean dip for fat-free tortilla chips or vegetable sticks. Of course, canned refried beans are a convenience item. You can make your own pretty easily at home. Just heat a can of pinto beans and mash them with a fork or bean masher. No need to add anything else. However, you could develop your own signature recipe: add some chopped onions, grated carrots, salsa, jalapeño peppers, garlic—you be the chef! If the fat-free beans seem a little dry, stir in some water a tablespoon at a time until the consistency is the way you like it.

There is also another convenient bean product that you might be interested in trying. The product is made by Taste Adventure and is simply called Pinto Bean Flakes. Some supermarkets may carry it with the natural foods brands; otherwise, you'll have to go to your neighborhood natural foods store. The bean flakes are precooked and instant—

it only takes five minutes to heat them with water in a microwave oven—and they turn out beautifully. So they're quick, and they save you from having to mash whole beans.

What about taco shells and those taco and burrito "kits"? Take a look at this label:

taco shells

Ingredients: corn treated with lime, partially hydrogenated soybean oil, salt	Serving size: 1 shell Calories: 55 Protein (gm): 0 Fat (gm): 3 Carbohydrate (gm): 6 Cholesterol (mg): 0 Sodium (mg): 50 Calcium (% USRDA): > 2 Iron (% USRDA): > 2

Taco and tostada shells (taco shells are curved like a pocket; tostada shells are flat) are made from corn tortillas that have been fried in oil. As you can see, each shell contains three grams of fat. If the taco filling and tostada topping are fat free (fat-free beans, chopped greens, tomatoes, onions, salsa . . .), then the total fat content of a couple of tacos or a tostada made with these shells may be fine.

On the other hand, you can save yourself some fat by using soft (not fried) fat-free flour or corn tortillas instead of the fried shells. Soft tacos are made with flour tortillas instead of taco shells. (Same filling, but the soft tacos are sometimes easier to hold and less messy to eat because they conform to the shape of your hand.) Flour tortillas can be found in the refrigerator case (see Aisle 12), or they come with the soft taco kits that are available here on Aisle 4. They're essentially fat free.

You'll also see boxed burrito kits. These kits call for ground beef, cheese, and tomatoes to be added at home. Use refried beans instead of beef, and if you use cheese, it should be nonfat. Also note that these kits often contain sulfites. Like the pizza kit described earlier, you can assemble tacos and burritos very easily and conveniently with individual ingredients instead of buying an all-inclusive kit.

Another word about cheese: some people like to use soy cheese instead of regular cheese. Soy cheeses are cholesterol free, but they are frequently high in fat. At the time of this writing, few, if any, are fat free. If you use soy cheese, look for a low-fat variety and use it as a garnish, rather than piling it on liberally. In most parts of the country, you will probably have to go to a natural foods store in order to find soy cheeses (read more about cheeses on Aisle 12).

Coffee, Tea, and Cocoa Mixes

Are you a coffee lover? Decaf or regular? Then there is the issue of flavor—chocolate almond,

---

Talking shopping with Robert Pritikin, director of the Pritikin Longevity Centers, Santa Monica, California:

I use fat-free cheese when I make tostadas—that's the only way I can eat fat-free cheese. I grate it when it's cold. You can't eat fat-free cheese plain—it's terrible—and you can't heat it—it just balls up. It's not good on pizza. You're better off just leaving the cheese off pizza altogether and loading it up with lots of vegetables.

About once a week at home, my wife makes a big batch of refried beans by cooking pinto beans with spices and then blending the mixture. We sometimes freeze some of that. Then, to make tostadas, we take fresh corn tortillas and bake them in the oven 'til they're crisp. We put some of the beans on top of the tortillas, add some grated nonfat cheese, then add salsa, lettuce, tomato, onions, and cilantro—you've got to have cilantro. Top all of that with some fat-free sour cream.

hazelnut, amaretto . . . Or maybe you are a fan of iced coffee?

Whether coffee is your beverage of choice or not, here are the java facts: research is unclear about the overall effects of caffeine on health. Caffeine is a stimulant found in coffee, tea, and cola drinks. Caffeine can cause you to have heart palpitations, stomach upset, and diarrhea (some people use it as a laxative!). It can also cause you to have the caffeine jitters and can disrupt sleep patterns. Some people are more sensitive to it than others.

Besides these effects, though, other health problems have not been linked to caffeine with much certainty. And some of these effects can be lessened by drinking decaffeinated coffee, although the small amount of caffeine that remains in decaffeinated coffee can still give some people trouble.

There is also a nutritional issue to consider. Coffee and tea—regular *or* decaffeinated—contain tannic acid. Tannic acid binds with minerals such as calcium and iron, making them less available to your body. In this way, coffee and tea can act as an antinutrient, robbing your body of nutrition. If you do drink coffee or tea, it is best to drink them between meals, rather than at the same time as meals. That way, the negative effects on your body's mineral absorption are lessened.

WHAT ABOUT COFFEE SUBSTITUTES?

Some instant hot grain-based beverages that are marketed as coffee substitutes have been available for many years. One popular brand is made with malted barley, barley, chicory, and rye. Another is made with bran, wheat, molasses, maltodextrin, and natural coffee flavor. My personal favorite is a product called Kaffree Roma, which is made by the Natural Touch division of Worthington Foods, Inc. It can be found in natural foods stores or Seventh-Day Adventist stores.

These coffee substitutes contain no caffeine and are perfectly fine for anyone who wants to use them. The flavors of various brands differ somewhat, so if you want to give this type of product a try, you may need to taste a few before you find one that you like. These products have a taste all their own, which bears a resemblance to that of coffee but is still quite different.

There are also some international instant coffees on the market that are flavored. Here is a sample ingredient label from one of them:

Viennese chocolate instant coffee

Ingredients: sugar, milk chocolate, nondairy creamer, instant coffee, cocoa, trisodium citrate, artificial flavor

This coffee has two grams of fat in a six-ounce cup.

As you can see from the label, this is a mixture of sweeteners, creamer, and coffee—not much different than if you simply added a little sugar, milk, and chocolate flavor to a cup of regular coffee at home. These coffees are not really better or worse for you than regular coffee with cream and sugar.

SPEAKING OF WHICH . . .

This brings us to the question of what you put *into* your coffee. And what's the bottom line on coffee anyway—is it an okay choice or not?

The drawbacks to drinking coffee have already been discussed. Other than that, any additional nutritional impact depends on what else you might add to it. Have you looked at the ingredient list on a container of powdered creamer lately? If not, you may be shocked at the length of the list. The first couple of ingredients are likely to be corn syrup solids and partially hydrogenated soybean oil—followed by a long line of additional ingredients, all of which are present in small amounts.

The fat content of a cup of coffee is usually about two grams, give or take a gram, whether you use real cream or a nondairy creamer. If you only drink one cup, then that's not a great deal of fat. But if you drink several cups of coffee per day, it adds up. A teaspoon of sugar in a cup of coffee is not much of a problem, either. True, you are better off drinking your coffee black. But the decision about whether or not to drink coffee should hinge more on the issues already discussed.

I recommend avoiding coffee. Coffee is not a good choice, primarily because of its role as an antinutrient and because of the side effects. If you do drink coffee, regular or decaf, try to hold it to two cups per day or less. That's the bottom line.

IS IT TEATIME?

The situation with tea is pretty much the same as that for coffee. There's less caffeine in tea than in coffee, but there is still a significant amount. Like coffee, tea is a potent antinutrient. It robs the body of minerals. An exception is herbal tea. If you enjoy herbal tea—peppermint, cinnamon spice, raspberry, orange, and so on—it can be an acceptable alternative to regular varieties.

COCOA MIXES AND POWDERED MILK

Chocolate does contain caffeine, but the amount in chocolate-flavored drinks isn't much. The fat content of chocolate syrups and cocoa mixes is not particularly high, either. A typical hot cocoa mix

---

HELPFUL HINT

If you are an avid iced-tea drinker and would like to cut back, try diluting the tea with another noncaffeinated beverage such as apple cider or lemonade. Better yet, combine these beverages with iced *herbal* tea.

contains one or two grams of fat per cup, and many are now fat free or contain no tropical oils (which are known to raise blood cholesterol levels).

Most hot cocoa mixes are best described as "nutritional neutrals," assuming that not more than a cup or two is consumed per day and that they are mixed with water, skim milk, or a low-fat soy milk. If more than that is consumed, then the calories from the beverage begin to displace calories that could otherwise come from more nutritious foods and foods with fiber. In other words, the hot cocoa begins to push better foods out of the diet. Some mixes are also flavored with artificial or alternate sweeteners such as aspartame. These are not recommended and are discussed in more detail in Aisle 7.

Finally, you'll also find nonfat dry milk here, as well as cans of goat's milk and cans of evaporated

GREAT CHOICES

- fat-free sloppy joe sauce
- pizza crust mix—top it with tomato sauce and veggies
- fat-free bottled spaghetti sauces
- fat-free canned tomato sauces, tomato puree, crushed tomatoes, and tomato paste
- egg-free dried or fresh pasta, or fresh pasta made with egg whites instead of whole eggs
- matzo
- borscht
- kosher fruit fillings and toppings—prune, cherry, red raspberry, apricot, wild blueberry, pineapple
- bean thread noodles
- fortune cookies
- canned refried beans, especially the fat-free ones
- flour tortillas
- herbal tea

and sweetened condensed milk for baking. Milk is discussed in detail in Aisle 12. Suffice it to say that when any dairy product is used, it should be nonfat or low-fat and limited to two cups per day (see the discussion of calcium needs in Nutrition in a Nutshell).

Talking shopping with Sally Silverstone, one of the original inhabitants of Biosphere II, Oracle, Arizona:

(I interviewed Sally in Houston, Texas, at the Johnson Space Center where we were both participating in a NASA workshop on the use of vegetarian diets in space travel.)

I always try, still, for my produce to go to a produce shop. So, it depends on time. If I have time to go to the produce shop, I look for all my week's produce in there. That's basically because the one I go to tries to stock locally grown food. These shops are dying out; we're lucky enough to still have one in the area where I work. So, I go to him first.

Now, often I don't have time to go to him, in which case when I go to the supermarket, I suppose the bulk of what I'm actually putting into the basket is fresh produce. I'll go and get the lettuces, tomatoes, peppers, cucumbers, carrots, squash for the week and what have you.

Now, the other things I buy are (and I'll try to buy them in bulk and locally, anyway) bags of beans, bags of brown rice, ready-made whole wheat tortillas—because I love to eat tortillas and I don't have time to make them myself, so I buy huge packets of tortillas. I buy all my herbs and spices (I don't have time to grow them anymore, although I used to). I probably buy one small carton of milk a week—because being English I have to have my early morning cup of tea and I have to have a drop of milk in it—I'll buy a cottage cheese and a yogurt a week.

Aisle 5

Cereal	Crackers
Oatmeal	Candy
Cookies	

Grocery carts slow to a crawl when they hit this aisle. Cookies and crackers are serious business, and there are so many choices now. This is also the aisle that brings out the coupons—a new breakfast cereal seems to hit the shelves every month. We'll take a look at the best choices in all of these, not to mention the breakfast drinks, granola bars, fruit roll-ups, and more.

Breakfast Cereals—Hot or Cold

First of all, cereals certainly aren't just for breakfast. A bowl of cereal can be a great choice as a snack or even a light supper. Sure, there are plenty of cereals that are hardly more than refined flour and sugar—more like confections than substantial, wholesome foods. And there are some that are "nutritional neutrals"—cereal's equivalent of white bread (it's not going to hurt you, but it's not a big asset to your diet either). But there are lots of great choices out there now, too, and we'll take a look at how to spot them.

DRY CEREALS

The very best dry cereals are made entirely or primarily with whole grains, with little or no added fat, salt, or sugar. There are only a few in the supermarket that actually fit that description. One example is shredded wheat. Read the ingredient list and what does it say? Nothing but 100 percent whole wheat.

But there are plenty of others that, while not being quite so pure, are still close enough. Compare these two labels:

sugar-coated flake cereal

Ingredients: corn, sugar, salt, malt flavoring, corn syrup	Serving size: ¾ C (cereal alone)
	Calories: 120
	Total fat (gm): 0
	Protein (gm): 1
	Carbohydrate (gm): 28
	Cholesterol (mg): 0
	Dietary fiber (gm): 0
	Sodium (mg): 210
	Calcium (% DV): 0
	Iron (% DV): 10

raisin bran cereal

Ingredients: whole grain wheat, raisins, wheat bran, sugar, natural flavoring, salt, corn syrup, honey	Serving size: ⅔ C (cereal alone)
	Calories: 120
	Total fat (gm): 1
	Protein (gm): 3
	Carbohydrate (gm): 31
	Cholesterol (mg): 0
	Dietary fiber (gm): 6
	Sodium (mg): 200
	Calcium (% USRDA): 0
	Iron (% USRDA): 35

The first cereal is primarily refined flour, sugar, and salt. It contains no fat and no cholesterol. But it also contains no fiber. It's a good example of "it won't hurt you, but it won't help you, either."

The second cereal is a different case. The first two ingredients are a whole grain and a whole fruit. Next is bran. These first three ingredients are the predominant ones in this cereal. Only after them come the sweeteners, flavoring, and salt. This cereal contains a nominal amount of fat (which naturally occurs in the whole wheat) and no cholesterol. It is also a significant source of fiber—six grams per serving. All things considered, this cereal is a great choice.

Here is another pair to consider. How would you evaluate these fruit-and-nut muesli cereals?

domestic fruit-and-nut muesli

Ingredients: raisins, whole wheat, whole barley, corn, wheat bran with other parts of wheat, rice, pecans, oats, sugar, brown sugar, corn syrup, dextrose, dried figs, honey, salt, partially hydrogenated vegetable oil, dried pears, sorbitol, modified food starch, natural flavors, dried blueberry puree, glycerin, blueberry juice concentrate, molasses, malt syrup, color additive (blue 1), citric acid, BHT; fruit preserved with sulfur dioxide	Serving size: ½ C (cereal alone) Calories: 150 Total Fat (gm): 3 Protein (gm): 4 Carbohydrate (gm): 30 Cholesterol (mg): 0 Dietary fiber (gm): 3 Sodium (mg): 150 Calcium (% USRDA): 0 Iron (% USRDA): 20

imported fruit-and-nut muesli

Ingredients: wheat bran, whole wheat and rye flakes, whole oat flakes, sugar, honey, hazelnuts, raisins, dried apple, almonds, wheat germ, salt, cinnamon	Serving size: ½ C (cereal alone) Calories: 140 Total fat (gm): 3 Protein (gm): 5 Carbohydrate (gm): 31 Cholesterol (mg): 0 Dietary fiber (gm): 7 Sodium (mg): 85 Calcium (% USRDA): 2 Iron (% USRDA): 15

These two products are similar in fat and calorie content, and neither contains any cholesterol (cholesterol is only found in animal products). The fat content is relatively low (despite the addition of nuts). Both ingredient lists show whole grains, dried fruit, nuts, sweeteners, and salt, but in slightly different orders. The imported cereal contains fewer total ingredients; the domestic cereal includes some preservatives and artificial color. The imported cereal also contains more than twice the fiber of the domestic brand.

Which would I choose—if either? I would take the imported muesli, even though it probably costs more. The domestic brand contains BHT and sulfur dioxide—preservatives that I prefer to avoid, if possible—as well as artificial coloring. Hydrogenated fat has also been added, which can raise blood cholesterol levels.

Both products contain nuts, which are best avoided or limited due to their fat content. However, in this case, the nuts in the imported cereal are really a minor ingredient, and the overall fat content of the cereal is relatively low at three grams per half-cup serving. This cereal could even be mixed with another, such as whole grain flakes, to dilute the nut factor even further.

Here's one more comparison—a look at regular granola and the new low-fat variety:

regular granola

Ingredients: rolled oats, brown sugar, partially hydrogenated sunflower oil, defatted wheat germ, high-fructose corn syrup, salt, molasses, natural flavor, lecithin	Serving size: 1 oz. (about ¼ C; cereal alone)
	Calories: 130
	Total fat (gm): 4
	Protein (gm): 3
	Carbohydrate (gm): 18
	Cholesterol (mg): 0
	Dietary fiber (gm): N/A
	Sodium (mg): 80
	Calcium (% USRDA): 0
	Iron (% USRDA): 6

low-fat granola

Ingredients: rolled oats, whole grain rolled wheat, brown sugar, raisins, crisp rice (rice, sugar, salt, barley malt), partially hydrogenated cottonseed oil, glycerin, nonfat dry milk, honey, dried coconut, almonds, salt, natural flavors, cinnamon, vanilla extract	Serving size: 1 oz. (cereal alone) Calories: 110 Total fat (gm): 2 Protein (gm): 3 Carbohydrate (gm): 22 Cholesterol (mg): 0 Dietary fiber (gm): 2 Sodium (mg): 50 Calcium (% USRDA): 2 Iron (% USRDA): 4

These two cereals are very similar in nutritional value. Both are made primarily with whole grains and sugar, with some cholesterol-raising, partially hydrogenated oil and other miscellaneous ingredients added in smaller amounts. The low-fat variety contains some dried coconut, which is high in saturated fat, as well as almonds, but apparently these ingredients are not enough to affect the fat content of the cereal significantly. At two grams of fat per serving, the low-fat granola has about half the fat of the regular granola.

However, note how small the serving size is for granola: one-fourth of a cup! That isn't very many spoonfuls. Eat enough of it, and even the amount of fat in the low-fat variety could add up.

I generally suggest avoiding foods that contain added fats, such as the hydrogenated oils added to these cereals. Of course, if the fat is present in low enough levels, it may be feasible to work products such as these into your diet. You will need to use your own judgment. Some people prefer following hard-and-fast rules and find it easier simply to avoid any product that contains added fat.

On the other hand, either of these cereals might be fine to include, depending on how much you ate. If you like granola, one way to eat it is to mix it with other cereals that are lower in fat. Since one quarter cup of cereal isn't much, you could mix a quarter cup of granola with a larger portion of another whole grain, low-fat cereal, such as bite-sized shredded wheat biscuits.

WHAT DO YOU PUT ON YOUR CEREAL?

Skim milk, if you use milk. But I'd like to know how we ever started using milk on cereal in the first place. The person who started that tradition obviously didn't speak with my grandfather.

My grandfather would be well over one hundred years old if he were alive today. He had his own mealtime traditions. He poured coffee on his cereal—never used milk. Watching him, I learned to keep an open mind about such things. Later, when I was in college and living in my first apartment, I poured orange juice over my cereal when I ran out of milk.

It may come as a shock to some that there are so many alternatives to milk for cereal. At first, the ideas just mentioned may sound pretty gruesome. But if you talk to people who put apple juice and orange juice on their cereal, they'll tell you they love it. I have a friend who uses applesauce on his

---

HELPFUL HINT

For a change of pace, try mixing two or more low-fat, whole-grain dry cereals together. For example, mix a small amount of low-fat granola with some raisin bran cereal, or mix a multigrain, fruit-and-nut-type cereal with spoon-sized shredded wheat biscuits. This works especially well when one of the cereals is a little higher in fat than the others, such as some granolas or cereals that contain nuts. Mixing these cereals with others that are lower in fat helps to bring the total fat content down.

cereal—just applesauce and nothing else. As for me, I like low-fat soy milk on my cereal. Applesauce with soy milk on cereal is good, too.

HOT CEREALS

These are fairly straightforward. The best hot cereals are whole grains. Rolled oats is a good example of an excellent choice. Others are made with whole wheat, whole rye, or blends of various whole grains. Hot cereals such as grits or the Cream of Wheat variety are refined and have very little fiber. They're another example of nutritional neutrals. Go with whole grains instead when you have the choice.

As for the individual packets of flavored oatmeal: these are generally fine. Some contain artificial flavors, but others contain natural flavors and spices. You could very easily flavor your own hot cereals at home by sprinkling a little cinnamon on top, tossing in a handful of chopped dried fruit, raisins, or currants, or adding some pieces of fresh fruit.

Breakfast Drinks and Specialty Bars

Next to the boxed cereals are the products for those who would otherwise skip breakfast: the meal-in-a-gulp instant-breakfast drinks and the meal-in-a-couple-of-bites breakfast bars. Actually, these products are marketed for "fitness," "weight control," and as "well-balanced meal replacements."

The nutritional composition of the canned drinks varies, and the amount of fat can range from 2.5 grams to 10 grams in a twelve-ounce can. Nonfat milk, water, sugar, and oil are typical first ingredients. The powdered-mix variety is much the same, although the ultimate nutritional value depends on whether it is mixed with skim, low-fat, or whole

Talking shopping with Jane Wiedlin, rock musician and former member of the Go-Gos, Valley Center, California:

My husband and I are big on carbo loading! The majority of my food purchases are bread, pasta, and rice items. We eat bagels or English muffins and cereal every day for breakfast. I use the fat-free cream cheese; my husband prefers the low-fat variety. I like using apple juice on my cereal. It sounds a bit strange but is really tasty and helps me in replacing dairy products, which I avoid whenever possible.

For lunch, we usually have soup (Progresso makes some delicious veggie soups that I'll 'doctor up' with spices and fresh vegetables) and/or sandwiches. Yves makes fat-free veggie hot dogs and a super veggie burger (Burger Burger) that we both enjoy.

Dinner is usually pasta (love it!) with either an Italian or Oriental twist. We also like to make those do-it-yourself pizzas. My favorite combo is veggie sausage, tomatoes, fresh chunks of garlic, broccoli, and *lots* of hot peppers! We are in no way 'fancy eaters,' and our choices, though vegetarian, are not always completely healthful. I do buy fat-free products when given a choice. The way I figure it, this makes it easy to maintain your weight without suffering!

milk. The breakfast bars are very similar to the granola bars described next.

These products are not particularly effective for any of the situations for which they are marketed. For "fitness," as a quick snack on the go, you'd be just as well off eating a bagel, fresh fruit, a bowl of cereal, or any number of equally quick or portable foods. As for "weight control" and as a "meal replacement," just how satisfying is a liquid meal? Not very. It's a good bet that if breakfast consisted of a flavored drink or a four-bite bar, it wouldn't be long before you'd be looking for something more to eat.

Granola Bars and Toaster Pastries

Granola bars are placed alongside the cereals and breakfast drinks, but they are probably as popular as snacks and lunch-box add-ons as they are for breakfast, if not more so. They're *small*—smaller than the pictures on the boxes. Literally four bites. And because they're made with cereal, they have a more healthful image than, say, a candy bar.

The ingredient list usually starts out, as expected, with the ingredients for granola. Some granola bars have extras added, such as peanut butter, chocolate chips, or pieces of caramel or fruit; some have chocolate coatings. So the fat content can vary from less than two grams per bar to over five grams. Fiber content is only about one gram per bar.

Granola bars are better than candy bars as a treat, and those that are lower in fat can be an acceptable pick-me-up or portable snack on occasion (look for those made without added oils). Watch out for the bars that are loaded with chocolate or peanut butter, though. These begin to approach the nutritional quality of some candy bars, especially if you don't stop at just one.

As for toaster pastries—what can be said for a breakfast item that is made of preserves, refined flour, fat, sugar, and artificial color and flavor? Not recommended.

If you really like toaster pastries, the natural foods brand variety that was mentioned in Aisle 1 is a better choice—made with whole grain flour and natural fruit flavorings.

Kiddie Fruit Snacks

What used to be called "fruit leather" has evolved. Now it's "fruit by the foot," "fruit roll-ups," and fruit in cartoon shapes. Some shapes are even action-packed, with liquid centers that gush—just what every parent wished for.

These products are made with fruit, sugar, artificial color, sulfites, mineral oil, and beeswax. They're certainly fat and cholesterol free (the mineral oil and beeswax just pass on through). But a far more wholesome choice, considering the other ingredients, would be a similar product found at a natural foods store and made with all fruit and no artificial color and preservatives. Some people even like to make their own fruit leather at home, using a fruit dehydrator. All of these stick to the teeth, however, so tooth brushing should be encouraged after eating.

Candy

A few acceptable candy choices were described in Aisle 2—the fat-free pectin jelly beans, fruit sours, and hard candies. You might wonder why these items are listed as great choices when other empty-calorie foods such as gelatin mixes and the kiddie fruit snacks discussed earlier are not.

This is how I differentiate among these choices: fat-free sweets that are meant to be eaten as condiments (such as preserves and jelly) or in small amounts and those meant to be eaten as treats (such as jelly beans and hard candies) are listed as great choices if they are also free of unwholesome ingredients such as artificial colors and flavors. There are plenty of them available, so I feel it's worth pointing them out. At some point, decisions like these may come down to a judgment call—this is part of the gray area of nutrition. One of my duties as your tour guide is to give you my educated opinion and to explain my rationale.

Candies like those on Aisle 2 are more expensive,

Talking shopping with Dr. John McDougall, physician and author of *McDougall's Medicine*, and Mary McDougall, author of *The McDougall Health-Supporting Cookbooks* (2 vols.), Santa Rosa, California:

(I spoke with John at his office in Santa Rosa and spoke with Mary later by phone. Mary is the family cook, and when she isn't home, John doesn't cook much. Instead, he keeps his meals very simple—a microwaved baked potato, a piece of fresh fruit, and so on. Here's what you'd usually find in Mary's grocery cart.)

frozen hash brown potatoes

pasta—egg free and oil free

oil-free pasta sauce (for times when she doesn't make her own and her youngest son wants to have pasta)

brown rice

dried beans, including pinto beans

barbecue sauces

canned garbanzo beans ("I love them. When I'm hungry, I just open a can and eat them plain, straight from the can. John would never do that—he hates them.")

pretzels

Tostitos tortilla chips ("Fat-free Baked Tostitos are the best of all of the chips.")

no-fat refried beans (again, for when the kids want something fast; otherwise she makes them from dried)

salsa

salad dressings—all types of oil-free dressings

gourmet-style candies, and these tend to be made with more natural colors and flavors than are the less expensive candies found on this aisle. If you don't mind gelatin and artificial color and flavor, then many of the hard candies and other fat-free candies on this aisle may be perfectly acceptable to you.

As for the chocolate candies, candy bars, and other goodies found on this aisle, none of them is recommended, but a couple of the better choices, if you must have chocolate, are chocolate-covered peppermint patties and chocolate-covered raisins, both of which are relatively low in fat. If you are someone who can eat just one piece of your very favorite candy or chocolate and be satisfied—without eating a pound of it—then go for it. One bite never hurt anyone.

Cookies

Cookies and crackers take up an entire aisle at my supermarket. The most striking observation I have made in browsing through this aisle—aside from the sheer volume of boxes and bags—is the total absence of cholesterol-raising lard and tropical oils (palm oil, coconut oil, cocoa butter) on the ingredient lists of most of the products. It almost seems as though manufacturers made this transition overnight.

Cookie manufacturers are now advertising their use of vegetable oils. The other noticeable trend is that they are removing much of the fat altogether. Lots of package fronts are being flagged with those magical words, "Fat Free" or "Low Fat."

However, if you haven't taken a stroll down the cookie aisle lately, then don't do it now. Even the new low-fat cookies are primarily refined flour, sugar, and fat. If you aren't already into the cookie habit, there's no reason to start. There are one or two exceptions. One brand, Health Valley, that is also sold nationally at natural foods stores, makes cookies from whole grain flour and fruit-juice sweeteners. These cookies are fat free and are a great choice. Fig Newtons are also a pretty good bet. Even though they contain the usual cookie ingredients—sugar, refined flour, and shortening—figs are the first ingredient, making these cookies a good source of fiber. Some Fig Newtons are now fat free, too, and these are the best choice. Again, more food companies are making very low-fat and fat-free cookies, and these are the ones to choose if you want a cookie. If you can find one that is made with whole grains, great. If not, then at least you can avoid the fat. Take a look at the two low-fat cookies below:

carrot walnut soft low-fat cookies

Ingredients: enriched flour, raisins, brown sugar, carrots, oatmeal, fructose, honey, partially hydrogenated vegetable shortening (soybean and cottonseed oils), walnuts, dextrose, leavening (baking soda, ammonium bicarbonate, cream of tartar), maltodextrin, egg whites, salt, soy lecithin, cinnamon, spice	Serving size: 1 cookie Calories: 60 Total fat (gm): 1 Protein (gm): 1 Carbohydrate (gm): 11 Cholesterol (mg): 0 Sodium (mg): 45 Calcium (% USRDA): 0 Iron (% USRDA): 2

animal crackers

Ingredients: enriched flour, high-fructose corn syrup, sugar, vegetable shortening (partially hydrogenated soybean oil), yellow corn flour, whey, baking soda, salt, artificial flavor	Serving size: 12 crackers Calories: 140 Total fat (gm): 4 Protein (gm): 2 Carbohydrate (gm): 23 Cholesterol (mg): 0 Sodium (mg): 160 Calcium (% DV): 0 Iron (% DV): 6

These two cookies are examples of decent low-fat options. They're basically nutritional neutrals, but at least they aren't loaded with the fat of, say, the typical double-chocolate chocolate chip cookie or any number of others. It would be better to find cookies made without the hydrogenated fat or any fat added at all. However, compared with almost all other cookies, these are a pretty good choice for every once in a while.

When you evaluate cookies like these, however, pay attention to the serving size. Some have one or two grams of fat per cookie. But if the cookie is bite-sized and you eat six or eight at a time, then the fat can add up pretty quickly. Another good example is the chocolate-flavored graham-type bear cookies—tiny animal-shaped cookies. There are two grams of fat in five pieces. I know plenty of people who will attest to eating half the box at one sitting. It adds up.

What about the imported cookies that have no nutrition labeling on their packages? Are the imported varieties of gingersnaps and animal crackers low in fat?

It's a tough call, and you'll have to rely on what you can deduce from ingredient lists and what you can tell subjectively from sampling the cookies themselves. If you are evaluating an imported brand of animal crackers or gingersnaps—varieties that are also made by domestic companies—compare the list of ingredients on the imported brand with those on the domestic brand, which has nutrition labeling. Where does the shortening fall in the list of ingredients? Does the cookie leave a greasy film on your fingers? Does it taste superrich and buttery? If so, don't buy these cookies again.

I noticed one domestic brand of shortbread cookies that was relatively low in fat. Imported Scottish shortbread cookies, on the other hand, are loaded with it. So, similar types of cookies are not necessarily similar in nutritional content. When in doubt, some people may be better off sticking to products with nutrition labeling or ones that contain no added fat or other fatty ingredients.

Talking shopping with Casey Kasem, radio and TV host, Los Angeles, California:

Gandhi said, "The only way you can break a bad habit is if it stands in the way of something you want more." That's the way that I maintain my lifestyle—I want to eat this way. My diet is under 20 percent fat—probably 10 to 15 percent fat.

CRACKERS

It's much the same story in the cracker department. Actually, we've seen some of the best cracker choices on previous aisles—the whole grain, low-fat crackers from natural foods companies on Aisle 1, the fat-free, whole grain flat breads on Aisle 2, and the matzos on Aisle 4 (page 149). Most other commercial crackers are like the cookies just described, except with less sugar. Most consist primarily of refined flours and shortening. Some are relatively low in fat, though, as compared to others. The rich, buttery-flavored crackers are usually higher in fat. Sometimes they leave a greasy film on your fingers—a telltale sign. Read labels to judge the relative fat content of crackers.

As with the cookies, one of the keys in choosing a cracker is to look at the serving size. If a serving of four crackers contains two grams of fat, that may be a good deal if the crackers are big. If they're minuscule and you are likely to eat a bunch, then think twice.

The Last Bite: What About Wheat Germ?

Wheat germ has been enjoying a healthy image for a long time. In the early seventies, I remember adding a couple of scoops to my morning health shake of bananas, milk, honey, and who-knows-what—the fix that was supposed to give me super-human energy so I could make it through my high school classes to my swim practices after school. The wheat germ was like fairy dust—it had special qualities, but I wasn't exactly sure what they were.

Now I have a better understanding. Wheat germ is a part of the wheat kernel that is separated out in the milling process. It's packed with vitamins and minerals, such as some B vitamins, vitamin E, and zinc. Three tablespoons of wheat germ (about one quarter cup) contain three grams of fiber, three grams of fat, and one hundred calories.

Is it recommended? Sure, but it's really a matter of personal preference. At a cost of three grams of fat and one hundred calories for a quarter-cup serving, wheat germ packs a lot of nutrition. Some people like the nutty taste it adds to hot or cold cereals, or they add it to baked goods or sprinkle it on top of casseroles or other baked dishes. Wheat germ is an example of a healthful, nutrient-dense food.

GREAT CHOICES

- low-fat, whole grain dry cereals, such as shredded wheat biscuits
- raisin bran, bran flakes, muesli, and others
- cereal toppers—low-fat soy milk, orange juice, apple juice, applesauce, skim milk
- oatmeal, including many individually packaged, flavored varieties
- whole wheat and rye hot cereals and other mixed whole grains
- fat-free fig bars such as Fig Newtons
- fat-free whole grain cookies
- fat-free whole grain crackers

Talking shopping with Kevin Nealon, comedian, *Saturday Night Live*, New York:

In my grocery cart, you would find:

baked tofu
fresh organic fruit, especially bananas
bagels
spaghetti
salad fixings
oatmeal
Rice Dream Original Lite [a rice-based milk
   substitute]
vegetarian refried beans, corn tortillas, soy cheese,
   and fresh salsa for soft tacos
Nayonnaise [a soy-based mayonnaise product]
chili beans, canned tomatoes, onion, and chili
   powder for chili
rice
sweet potatoes
artichokes
Smart Dogs or veggie wieners by Yves
Campbell's Vegetarian Vegetable Soup
soy pizza
Cedarlane hummus

Aisle 6

Baby Food
Canned Juices
Bottled Juices

Aisle 6, where juices predominate, is a relative breeze. Baby foods, too, can be found here. So, first a word about good nutrition for kids . . .

Are Low-fat Diets Kid Stuff?

The information contained in this book is primarily aimed at adults. Infants, children, and teenagers have some special nutritional needs, and to discuss these in detail is beyond the scope of this book. However, the nonprofit Vegetarian Resource Group in Baltimore, Maryland, has some excellent nutrition education materials that address some of the questions that you may have about adapting a low-fat, plant-based diet for the entire family. The address and phone number for the Vegetarian Resource Group are given at the back of this book.

In general, the special nutritional needs of young people center on nutrients that aid their rapid growth and development. This doesn't mean, however, that a low-fat vegetarian-style diet is not appropriate for young people. On the contrary, not only is this a healthful way of eating but the earlier a person adopts this lifestyle, the more likely it is that these good eating habits will stay with him or her throughout life.

But very low-fat diets are not appropriate for children under the age of two years. At that young age, children should be receiving breast milk or formula, or they should be graduating to table foods. This is not the time for actively restricting fat or being unduly concerned about a child's fat intake.

However, previous recommendations by leading health organizations for children over the age of two years stipulated that children should not be fed a low-fat diet. As I've mentioned, most nutritionists currently recommend a diet of about 30 percent fat for children. However, even this thinking is changing, and more health experts are beginning to feel comfortable with diets for children that range from 20 to 25 percent fat. Once again, it's worth noting that children in other cultures the world over have fared very well for generations on diets much lower in fat than this.

For young people, the nutritional issues involve ensuring that enough calories are consumed and that certain key nutrients, such as calcium, iron, and zinc, are provided in adequate quantities. Enough of these nutrients can easily be provided on a low-fat, plant-based diet, but limiting junk foods and paying attention to some basic nutritional principles would help increase the probability that these needs will be met.

With all of this in mind, most parents should be able to accommodate all family members on a low-fat vegetarian diet. Again, the Vegetarian Resource Group has materials that provide meal-planning guides, sample menus, and additional information specifically addressing the needs of the younger set.

Sports Drinks

You've seen them, and maybe you've used them. Chances are, you didn't actually need them.

If you drink sports beverages just because you like the way they taste, that's fine. They contain no caffeine, and some people like the fact that they are not carbonated. I suggest avoiding those that contain the nonnutritive sweetener saccharin, since it has been shown to be carcinogenic.

For those who are interested, the truth about fluid replacement during exercise is that for most people, plain water is the best beverage before, during, and after exercise. It's a good idea to drink a cup of water or two before you exercise, and to drink more—a cup or so—during exercise, especially in hot temperatures. Plain water is perfectly adequate and is beneficial for exercise lasting an hour or less.

As for sports drinks, there is some research that supports the use of sports drinks in cases where an athlete is exercising for an hour or longer at a time. The small amounts of carbohydrate and sodium in the drinks may help the body to absorb the fluid faster. Soft drinks and fruit juices are not as effective because these have a higher concentration of carbohydrates and thus take longer to absorb.*

At any rate, it is also notable that endurance athletes who consume a low-fat vegetarian diet that is at least 60 percent to 80 percent carbohydrate and not more than 20 percent fat show improved ath-

---

*Sports drinks vary in their composition. Some actually contain too much carbohydrate to be effective for athletes. For more help in choosing a sports drink, you may wish to contact a registered dietitian who specializes in sports nutrition. To find a sports nutritionist in your area, call the American Dietetic Association at 1-800-366-1655.

letic performance and greater endurance than those who eat a diet higher in fat.

Fruit Juice and Close Relatives

Some juices only *look* like fruit juice. They may taste like orange or cherry or guava, but this is really just a clever disguise. Fruit drinks are usually nothing more than sugar, artificial flavors and colors, and—oh, yes—10 percent fruit juice. Some are even enriched with vitamin C. Here's the label from a popular fruit punch, for example:

popular fruit punch

Ingredients: water, high-fructose corn syrup, 2 percent or less of the following: concentrated fruit juices (pineapple, apple, passion fruit, and orange), purees (apricot, papaya, and guava), citric acid, malic acid, calcium carbonate, natural and artificial flavors, dextrin, sugar, pectin, gum acacia, glycerol ester of wood rosin, red dye number 40, blue number 1, ethyl maltol, sodium benzoate, ascorbic acid	Serving size: 8 oz. Calories: 110 Total fat (gm): 0 Protein (gm): 0 Carbohydrate (gm): 28 Sodium (mg): 30 Calcium (% USRDA): 15 Vitamin C (% USRDA): 100

Another brand, whose juice is labeled "natural," contains water, sugar, fruit juice (10 percent), and natural flavor—a much better ingredient list, but it's mostly sugar water just the same.

Whether you buy bottled or canned fruit juice drinks or a powdered mix, these products are essentially sugar water. Like the sports drinks just described, at least juice drinks are caffeine free. Of course, they're fat free, too. They are fine for an occasional beverage. If artificial color and flavor

---
HELPFUL HINT

For a refreshing fruit-juice cocktail, pour a small amount of your favorite fruit juice into a tall glass. Fill the glass up the rest of the way with seltzer water or sparkling mineral water. Add a wedge of lemon or a twist of lime.

---

bother you, then stick with the "natural" varieties—but read the labels to be sure of what that means.

Other fruit-juice beverages are 100 percent juice with sugar added. Some contain saccharin, and some are sweetened with aspartame instead. Use of these sweeteners is a personal decision—I prefer to avoid them. Of all of the juice beverages from which to choose, the best choices are those that are 100 percent juice, with little or no sweeteners added and without artificial flavors and colorings. There are plenty of them, and some are now available as blends of two or more juices.

Vegetable juices are good choices as well. Tomato juice and vegetable-juice cocktails are high in sodium, though, so if salt is a problem for you, choose a low-sodium variety. They're readily available. Add a squeeze of fresh lemon juice or a shot of hot sauce to spice up low-sodium tomato juice, since it tends to taste flat compared to the regular salty variety.

FOR THAT SPECIAL OCCASION ...

Sparkling ciders can be a festive alternative to champagne for those who prefer to avoid the alcohol. Some are 100 percent apple juice (cider), some contain carbonated grape juice mixed with apple juice and vitamin C, and some are a combination of sparkling red and white grape juices. Like pure fruit juices, these are also a great choice.

The Last Bite: How Do Chocolate-Flavored Milk Drinks Rate?

Chocolate-flavored milk drinks are sold in small, aseptically packaged boxes, just like the fruit juices that are so popular for kids' lunch boxes. They also come in small bottles.

Here's a sample label from one container of chocolate-flavored milk:

chocolate-flavored milk

Ingredients: skim milk, water, sugar, cocoa, partially hydrogenated sunflower oil, corn syrup solids, salt, carrageenan, mono- and diglycerides, vanillin	Serving size: 8 oz.
	Calories: 150
	Total fat (gm): 2
	Protein (gm): 5
	Carbohydrate (gm): 28
	Cholesterol (mg): 0
	Sodium (mg): 85
	Calcium (% USRDA): 20
	Iron (% USRDA): 0

This product is relatively low in fat, but it does contain some added fats in the form of partially hydrogenated oil and mono- and diglycerides. Once again, some people may prefer to avoid foods that contain any added fat at all. If it makes decision making easier for you to apply strict criteria across the board, then that's fine. However, the

---

GREAT CHOICES

- 100 percent fruit juices—apple, pineapple, grapefruit, prune, blends, and so on
- tomato juice
- vegetable-juice cocktail
- 100 percent carrot juice
- sparkling cider
- sparkling grape juice

Talking shopping with Reed Mangels, registered dietitian and nutrition adviser, the Vegetarian Resource Group, Amherst, Massachusetts:

Here is what I shop for at the regular supermarket:

carrots, potatoes, and other vegetables in season
fruits in season
whole grain bread, whole grain English muffins
cold cereals—Cheerios, Shredded Wheat and Bran, Grape-Nuts
oatmeal (for breakfast and baking)
canned beans—garbanzos, kidney, pinto, and black beans
pasta
brown rice
low-fat spaghetti sauce without cheese or meat
vegetarian baked beans
tofu
rice cakes (Mother's brand)
pretzels
canned fruits
juice packs (winter dessert)
frozen peas and spinach
tomato sauce, paste, puree, and whole tomatoes
frozen, calcium-fortified orange juice
Pace picante salsa (good on baked potatoes, salad, and Mexican foods)
Sorrell Ridge Fruit Spread
unsweetened applesauce (replaces fat in baking)
whole wheat flour, spices, raisins, mustard, and ketchup

This is supplemented with produce from the farmer's market in season and trips to the natural foods store and co-op for tofu hot dogs, soy milk, organic fruits and vegetables, dried papaya, dates, figs (treats), grains (millet, couscous, quinoa), dried beans, soy yogurt, Ener-G Egg Replacer, and low-sodium soy sauce.

total fat content of this product is low enough that it can easily be worked into the diet.

On the positive side, this product contains no artificial colors or other objectionable ingredients. Vanillin is a synthetic but harmless substitute for vanilla flavoring. If a child wants to drink milk, this product is acceptable. Read labels, though. Some similar drinks contain artificial flavors and colorings or may contain more fat.

Aisle 7

```
Vitamins          Spices
Nuts              Shortening
Cake Mix
```

The Vitamins and Minerals Game

We've already said that there are many gray areas in the field of nutrition. Well, things really get murky when we touch on the subject of vitamin and mineral supplements.

A full-scale discussion of supplements would merit a volume of its own and is beyond the scope of this book. Your supermarket is likely to carry a variety of multivitamin supplements, children's multivitamins, and a few varieties of some of the most popular single nutrients, such as calcium and vitamin C. A drugstore is the place to go for a really extensive selection.

But the big question remains: should you take supplements? The experts are split on this one. Some very well-respected, well-qualified nutrition scientists take vitamin and mineral supplements. Especially popular at the present time are supplements of antioxidant nutrients, such as vitamin E, vitamin C, and beta-carotene.

Other equally reputable nutrition scientists feel

that there isn't enough evidence to merit the taking of supplements and that we should all rely on whole foods for what we need. Some make the case that nobody should rely on supplements to ensure a good diet, because there may well be substances that we need from our diet that have not been identified yet and that are not yet contained in multivitamin formulations, to give just one example.

Also, nutrients interact with each other. If we isolate individual nutrients from the diet—vitamins and minerals—and take them in high doses, we can upset the natural balance of nutrients in the diet, creating a problem where there wasn't one in the first place. For instance, zinc binds with copper. People who take large supplements of zinc can deplete their reserves of copper.

In any case, megadoses of vitamins and minerals—the high-potency, stress-formula supplements that are so common today—are not recommended for most people. They typically provide many, many times the Recommended Dietary Allowance (RDA) for the vitamins and minerals that they contain. The human body will simply flush out some excess nutrients, such as vitamin C or the B vitamins.

Other nutrients, however, are stored by the body, and high doses are not a good idea. If you do take a supplement, it's better to make it truly a *supplement* to your diet. In that case, you certainly do not need more than 100 percent of the RDA for any nutrient. You might even consider taking a lower-dose (regular-strength) vitamin-mineral supplement a couple of times a week, instead of taking one every day. Remember, as with so many things in life, more is not necessarily better. Do your best to get the nutrients you need from whole foods.

Nutty Ideas

A good-sized portion of this aisle is devoted to nuts of various kinds—cashews, walnuts, mixed nuts, peanuts—and in various forms: slivered, chopped, whole, salted, unsalted, and so on. Some are honey-roasted and others are coated with candy. Some actually *are* candy—such as the peanut-butter-and-chocolate peanut-shaped candies and other nut-and-chocolate combinations.

What do all of these have in common? Fat. Nuts are almost all fat; the nut-and-chocolate candies are mostly fat, too. No great choices here. Just move on.

Shortening

Fat is fat is fat, whether its name is canola oil, sunflower oil, corn oil, or olive oil. If you add fat to foods, add it sparingly. The idea is to limit your total intake of fat, no matter what the source.

The fat hierarchy, which dictates that canola oil is best, followed by safflower oil, then corn oil, then olive oil, with cholesterol-raising tropical oils, hydrogenated (solid) vegetable shortenings, and animal fats last, is essentially irrelevant if your fat intake is low enough. If fat makes up only 10 to 15 percent of the calories in your diet, then it makes little difference in the scheme of things

---

HELPFUL HINT

In some recipes (such as casseroles and salads) that call for chopped nuts, you can substitute chopped celery or chopped water chestnuts. They add crunch without adding fat.

---

whether that fat is canola oil or olive oil. At fat levels that low, there would be very little use of fat, and most of the fat in the diet would simply be inherent in the foods that you eat—the little bit of fat that is present in grains and vegetables, for instance.

The relative superiority of one oil over another takes on more significance when fat intakes are higher. How much significance is very difficult to say. Your best bet is to aim for a low total fat intake. Use vegetable-oil cooking sprays, which cover pans and baking sheets with only a thin film of fat, instead of slathering on shortening (buy those labeled chlorofluorocarbon free, or CFC free—they don't hurt the ozone). Sauté vegetables in a small amount of water or vegetable broth instead of using fat. Use flavored vinegars, balsamic vinegar, or lemon juice on salads instead of fatty dressings. See the following Helpful Hints for more ideas of ways to cut back on shortenings.

HELPFUL HINTS

- Invest in some good nonstick cookware—pots, pans, griddles, and so on. You'll be able to brown or sauté foods adding little or no oil to the pan.
- Infused oils are oils that have been flavored with lemon, garlic, hot pepper, or other distinctive seasonings. If you do use a drop or two of oil on occasion, these tasty oils will add lots of flavor for the amount used.
- It's difficult to judge how much fat there is in a splash of oil. Use measuring spoons and cups to measure fats in recipes accurately so that you don't fool yourself into thinking you are using less than you are.

Sugar and Spice

Playing with spices can bring out the artiste in anyone. The personality of a dish can be totally transformed with just a dash of the right ingredient. A pan of chopped vegetables sautéed with a little curry powder speaks a whole different language than the same vegetables cooked with fresh basil, for instance. Experiment a little. Most spices qualify as great choices.

If you don't feel confident using individual spices, try some of the spice blends or herb mixtures that come in shakers. Many are salt free, for those who prefer to limit their salt intake. Steer clear of flavor enhancers that contain MSG if you feel you are sensitive to this ingredient. Read labels.

BUTTER SUBSTITUTES

Butter substitutes that are packaged in shakers are also found here, and many of these are great choices. These are powdered, natural butter-flavored sprinkles that are fat and cholesterol free. They are made with natural flavors and colorings; they contain some salt, too. Some are sour cream or cheese flavored.

Some people may be surprised to see butter or other fatty-sounding ingredients listed on the label, but enough of the fat has been removed from these ingredients to render the product essentially fat free. Considering that products such as these are meant to be used as condiments, the risk of getting too much fat from them is small.

These products work well on hot moist foods such as steamed vegetables. Granted, I typically shun substitute-type products, particularly those made with artificial sweeteners or fat substitutes. In this case, however, the ingredients are benign; this particular kind of product can be very helpful for people who would otherwise have a hard time getting used to eating potatoes, rice, and vegetables without a little margarine melted on top.

SALT

Salt helps to bring out the flavor of foods, but as a culture, we tend to overuse it. It would be best if all of us reduced our salt intake.

Table salt is a compound made of sodium and chloride. About a quarter of the people who have hypertension are sodium sensitive and should limit their salt and sodium intakes to help control their blood pressure. And some people who have kidney disease or congestive heart failure also need to limit their sodium intakes. For the rest of us, though, it is probably fine to use a small amount of salt to flavor foods. A little salt can make many foods more palatable (although as many people can also attest, it doesn't take long to retrain your palate to like the taste of less salty foods).

All things considered, it's much more important for most people to reduce their fat and cholesterol intakes than it is to put much effort into decreasing the intake of salt.

On the shelf next to the regular table salt, you'll find salt substitute and something called "light" salt. Are these better than using regular table salt?

Not necessarily. Table salt is sodium chloride. So-called salt substitute is actually a different form

---

Talking shopping with Robert Pritikin, director of the Pritikin Longevity Centers, Santa Monica, California:

Here's a hot tip: buy the low-sodium V–8 juice and add a little Tabasco sauce to it. I always add a little Thai hot sauce—particularly if I'm eating a low-sodium soup—or a squeeze of fresh lemon. It brings out the flavor without adding sodium.

of salt—potassium chloride. Some people who are sodium sensitive like to use this product instead of regular salt (although there are other people who have to avoid the potassium for medical reasons). A drawback is that the potassium gives salt substitutes a bitter aftertaste.

Light salt is usually a mixture of regular salt (sodium chloride) and salt substitute (potassium chloride). If you do not have a medical reason for cutting back on regular table salt, I recommend continuing to use it (if you want to use any salt at all)—just use it sparingly.

SUGAR

Like salt, a little sugar doesn't hurt. The main problem with sugar is that it is associated with tooth decay, especially when sweet, sticky foods that adhere to the teeth are eaten and when oral hygiene is not maintained. (I can still hear my childhood dentist reminding me to "brush and floss, and when you can't, then swish and swallow.") The other problem is that many sweets also happen to be high in fat (pastries, cookies, ice cream, chocolate bars, and so on). Eating too many sweet junk foods can also push more nutritious foods out of the diet.

However, a little sugar—in whatever form—is not a problem for most people, as long as the diet is otherwise nutritious. What about artificial sweeteners? As I have already mentioned, I am not in favor of using nutritive or nonnutritive alternative sweeteners, such as aspartame and saccharin (both are marketed under several brand names), and I avoid them whenever I can. Saccharin is a weak carcinogen, and many case studies reported problems associated with the consumption of aspartame. Why bother with these, when a little sugar tastes good and is safe?

Assorted Baking Ingredients

A quick mention of baking ingredients such as baking soda, baking powder, yeast, and cornstarch: these are not a problem. Cocoa for baking is low in fat, too. It's the chocolate chips, the coconut, the butterscotch chips, and the white and dark chocolate almond bark that are problematic. The cookies and candies in which these are used are typically high in fat. And if there's a bag of chocolate chips in the cupboard, you might find yourself munching on them for a snack, instead of an apple. When you bake treats such as cookies and cakes, there are many ways to modify recipes to cut down the fat and improve the nutritional content. Take a look at some of the resources listed at the end of this book.

Bread Crumbs and Stuffing Mixes

It's not the bread crumbs that are the problem—it's what you do with them. Breaded foods are usually fried, like eggplant Parmesan. (Some bread crumbs contain MSG, too.)

Stuffing mix is a similar story. Most people associate it with turkey, which you already know is not recommended on a health-supporting, low-fat diet of 10 to 15 percent of calories from fat.

What about using it to make stuffed squash, or just eating the stuffing by itself? you might ask.

Like the bread crumbs, the stuffing mix itself is usually low in fat. (Some of the mixes contain sulfites, though.) But take a look at the instructions for preparation. One package that I saw stated that there was a mere gram of fat in one ounce of stuffing mix. But "as prepared"—adding a whole stick of margarine (eight tablespoons of margarine for eight cups of stuffing)—the end product contained about seven or eight grams of fat per half-cup serving. That's quite a bit of fat.

If you use stuffing mixes, cut the fat by at least half. You may need to compensate by adding more moisture in the form of extra fluid (vegetable broth, for instance), chopped fruits, or vegetables. Then hold your serving to a half cup and fill up on veggies and other good foods that have no fat added.

Bread Mixes and Muffin Mixes

Here's a list of the major ingredients in one typical supermarket muffin mix: white flour, sugar, vegetable shortening, artificial coloring, artificial flavoring, and artificial blueberries. Yuck. Most labels give no nutrition information.

Now take a look at the following bran muffin mix. The picture on the package label shows a dark brown, healthy-looking muffin—with a pat of margarine melting on one half. The package front claims, "Made with 100 percent vegetable shortening" and "With high fiber added," but no nutrition information is given.

bran muffin mix with high fiber added

Ingredients: bleached wheat flour enriched with vitamins (niacin, thiamine mononitrate, and riboflavin), sugar, vegetable shortening (partially hydrogenated soybean and cottonseed oils), wheat bran, dehydrated molasses, leavening (baking powder containing baking soda, sodium aluminum phosphate), salt, whey, dextrose

This mix contains the usual major ingredients: white flour, sugar, and shortening. Wheat bran is added, but we don't know how much fiber the muffin contains. The instructions on the package call for an egg and milk to be added. Not particularly healthy. Many of the bread and muffin mixes call for eggs and milk to be added. This is an Amer-

> **HELPFUL HINT**
>
> Next to the muffin mixes, you'll see the bags of flour. Among the white flours, you'll probably also find whole wheat flour. You might consider making muffins from scratch at home—they're quick and easy, and they freeze well. Almost any recipe can be made using half white and half whole wheat flour. You can even try making them using all whole wheat flour. Make an extra batch while you're at it, and freeze it for later.

ican cultural tradition—can you imagine any American-style cookbook *not* calling for eggs or milk in a cookie, cake, or bread recipe? As you know by now, many other ingredients—most lower in total fat, saturated fat, and cholesterol—can be used in place of these animal products (see The Produce Section, Aisle 1, and Aisle 12). Wouldn't it seem strange, though, to look up a recipe for pancakes in your Betty Crocker cookbook and find that it calls for a mashed banana instead of an egg?

Some mixes still contain beef fat and lard. Avoid them.

Having Your Cake . . .

Even following the "low-cholesterol" instructions on the box of cake mix, one slice of cake still packs about ten grams of fat. The ingredients? The usual suspects—sugar, white flour, shortening, artificial flavor and color. Some new light cake mixes contain three or four grams of fat per slice. The ingredients? More of the same: sugar, white flour, shortening, artificial flavor and color—no better than their fattier siblings.

If it's cake you want, one mix stands out. It's angel food cake. Don't laugh. Angel food cake is fat free, and its ingredients are relatively wholesome. If artificial flavor is listed, it's probably vanillin, which is one artificial flavor that is considered to be safe.

Plain angel food cake can be made more interesting by cutting it into chunks and tossing it with fruit salad; serving it in slices topped with fresh berries; or even mixing some cocoa powder into the batter for a chocolaty treat. We'll list angel food cake as a great choice—for a cake.

AND FROSTINGS?

Canned frostings from the supermarket are a chemical soup. The ingredient list on one can starts out, "Sugar, vegetable shortening, artificial color and flavor," then continues with a long line of chemical additives. Two tablespoons of the frosting—about enough to cover a cupcake—contain five grams of fat. A light version only contains one gram of fat in two tablespoons, but the ingredient listing is similar and contains a litany of chemical additives.

Pass on the canned frostings, and sprinkle powdered sugar on top of your cakes instead. Try topping a cake with a thin layer of apple butter or your favorite fruit spread. Delicious.

Cookie Mixes

It's much the same story here as it was for the other mixes on this aisle. Most, if not all, of these cookie mixes are high in fat, You're much better off buying low-fat cookies (see Aisle 5) or baking some from scratch.

If you do buy a mix, however, don't be misled by confusing package information. One name-brand brownie mix, for instance, appears to be low in fat at first glance—only two grams of fat per serving. Actually, the mix itself *is* low in fat. But "as prepared," one brownie contains nine grams of fat—hardly a low-fat food. The way the nutrition information is presented graphically on the package—the mix alone, then "as prepared"—it would be easy to think that there is less fat per brownie than there really is. When you read labels on mixes, be sure that you see the "as prepared" information before you buy.

This particular mix also calls for two eggs to be added, which adds cholesterol to the final product. Regardless, the ingredient list sure looks familiar—sugar, enriched flour, partially hydrogenated vegetable oil. . . . We've seen a lot of that combination on this aisle. There's a light brownie mix, too. It's lower in fat. The ingredients list is similar, but it contains more additives. My recommendation is to enjoy a small brownie—any brownie—now and then (infrequently) but generally to weed these low-fiber, usually high-fat sweets out of your life.

Pie Crusts

First of all, if you are going to have pie, the best filling choices are fruit fillings, which are low in fat and might even have a smidgen of fiber, or a low-fat, cholesterol-free and vitamin-packed pumpkin filling.* In terms of the crust (if you really need one), graham cracker is probably the best choice. But buy the bags of graham cracker crumbs, rather than buying a ready-made graham cracker crust, and make your own. Cut back on the amount of margarine called for on the bag. Ready-made graham

---

*Be sure to use only nonfat milk or soy milk when pumpkin pie recipes call for milk to be used. If a recipe calls for eggs, use only egg whites or another substitute. My own favorite pumpkin pie recipe uses soy milk and tofu instead of the traditional eggs and milk. It tastes great.

Talking shopping with Sally Silverstone, one of the original inhabitants of Biosphere II, Oracle, Arizona:

SH: Are you actively trying to eat a low-fat, plant-based diet?

SS: Yes, I am. Because since coming out of the Biosphere, I've seen the benefits of that so greatly. Also, when I came out of the Biosphere, I experienced a very rapid weight increase, which quite frightened me.

SH: How did your diet change coming out of the Biosphere?

SS: Well, I ate a lot more, and I went a bit crazy, you know. I had lots of treats and people offered to take me out for meals. There were all these temptations around that I hadn't had for two years, and I was shocked how quickly my weight went back up. So, after a month or so, I thought, Whoa, there. I'd better do something about this. Now I will still go out and have the occasional binge. I'll go out maybe once in a couple of weeks and have a big dinner and I'll eat what I like. I don't eat red meat at all anymore.

SH: What was your diet like in the Biosphere?

SS: It was very low fat, it was calorie restricted, it was no refined sugars at all, it was high fiber and nutrient dense—not totally vegetarian, although the amount of actual meat in the diet was minute.

SH: Do you know what percent of calories came from fat?

SS: I believe about 9 or 10 percent. Very low fat, very low fat. I think we ate, on average, about thirty grams of fat a day. Calories are about two thousand a day. So, in fact, we were probably a little higher than that—maybe 11 or 12 [percent fat]. There were times when our fat intake was even lower—it was about twenty-seven, twenty-eight grams.

cracker crusts contain five grams of fat per slice, and they aren't all that much easier to work with than a lower-fat crumb crust made from scratch.

Pastry pie crusts made from a mix are high in fat—about eight grams per slice—and they don't hold up all that well if you cut down on the fat. So here's a better idea: consider making a crustless pie. For instance, make the pumpkin pie filling, but instead of pouring it into a crust, pour it directly into a casserole dish instead (okay, so it's more like pumpkin pudding). Fruit crumbles or crisps, such as apple crisp or peach crumble, are fruit fillings topped with a crumbly oatmeal mixture on top. They're a good alternative to pastry or graham cracker crusts, but go easy on any fat that you might add.

Baking Mixes

Have you ever used an all-purpose baking mix to make pancakes or waffles? Take a look at the labels from these two products—one regular-style mix and one reduced-fat variety:

regular baking mix

One cup makes six pancakes (after adding milk and egg). Ingredients: enriched bleached flour, iron, thiamin mononitrate [vitamin B_1], riboflavin [vitamin B_2], vegetable shortening (one or both of the following partially hydrogenated oils: soybean, cottonseed), leavening (baking soda, sodium aluminum phosphate, monocalcium phosphate), whey, salt, dried buttermilk

Serving size: $1/3$ C
Calories: 140
Protein (gm): 3
Total fat (gm): 6
Carbohydrate (gm): 26
Dietary fiber (gm): < 1
Cholesterol (mg): 0
Sodium (mg): 140
Calcium (% DV): 6
Iron (% DV): 6

reduced-fat baking mix

Contains 50 percent less fat than the original baking mix	Serving size: 1/3 C
	Calories: 140
Ingredients: enriched bleached flour, iron, thiamin mononitrate [vitamin B₁], riboflavin [vitamin B₂], vegetable shortening (one or both of the following partially hydrogenated oils: soybean, cottonseed), leavening (baking soda, sodium aluminum phosphate, monocalcium phosphate), whey, dextrose, salt, buttermilk	Protein (gm): 3
	Total fat (gm): 2.5
	Carbohydrate (gm): 27
	Dietary fiber (gm): < 1
	Cholesterol (mg): 0
	Sodium (mg): 470
	Calcium (% DV): 4
	Iron (% DV): 8

Both mixes are primarily white flour and shortening, with milk and eggs added at home. The nutrition label on the first mix says that one-third of a cup of mix contains six grams of fat and no cholesterol. Of course, that's before the milk and eggs are added, but the label doesn't make that clear. The reduced-fat mix contains less fat than the regular variety, but it still calls for milk and eggs to be added. The package also gives instructions for cutting the cholesterol by using skim milk instead of whole milk and using egg substitute instead of whole eggs. (Of course, you could also use one of the egg substitutes that have been mentioned previously—mashed banana, tofu, and others.)

GREAT CHOICES

- most herbs and spices—basil, bay leaves, curry powder, coriander, cinnamon, cumin, dill, garlic, ginger, mint, paprika, pepper, rosemary, sage, thyme, tarragon, and so on
- natural, butter-flavored sprinkles
- angel food cake
- low-fat graham cracker pie crusts

If you use a baking mix, pick the reduced-fat variety and use skim milk or low-fat soy milk and mashed bananas or another egg substitute in place of whole eggs. A whole grain mix would be a better choice, but the white-flour type is still acceptable.

Talking shopping with Mary McDougall, author of *The McDougall Health-Supporting Cookbooks* (2 vols.), Santa Rosa, California:

In addition to shopping at the regular supermarket, I always shop at a natural foods store—mainly because of all the fat-free convenience products that you can get there. If I made everything from scratch, I could shop strictly at a regular supermarket and be fine. But the reality is that people are very busy today, and they like to use some convenience foods. It was fun to cook when I had five people to cook for. Now that it's only three, it's not as much fun.

One of the main things I buy at the natural foods store is instant soup cups. My eleven-year-old son likes them, and I like them. They're great for topping a baked potato—especially the rice-and-beans variety. You just heat the soup and let it set for a few minutes until it gets thick. Then pour it over the baked potato.

I also buy oil-free enchilada sauce for a snack for my eleven-year-old. He eats refried beans in a flour tortilla with enchilada sauce poured over the top. He heats it in the microwave.

My advice for people trying to make this lifestyle change is to eat lots of starch. Buy lots of pasta and grains. You don't want to be hungry, and these things fill you up. Eat lots of potatoes, rice, pasta, and beans. We snack on pretzels and bagels. I buy my bagels from the bagel store where I can buy them fresh, and they're better than what you buy at the supermarket.

The Last Bite: Liquid Spices

Some common spices—cinnamon, garlic, and oregano, for instance—can now be found in "liquid" form. Actually, these are tiny bottles of soy oil infused with the spice essence. In cooking or baking, the amount used is the same for the liquid spices as for dried spices—if the recipe calls for a half teaspoon of ginger, use a half teaspoon of the liquid spice. Liquid or dried spices are four times the strength of fresh spices.

A couple of teaspoons of liquid spice added to a batch of food may not add a significant amount of fat per serving. But every little bit adds up, and one teaspoon of oil contains five grams of fat. I recommend using the old-fashioned dried spices, or fresh ones.

Aisle 8

Soft Drinks
Mineral Water
Beer and Wine

Is red wine good for my health?

A few words about alcohol are due here. Then we'll zip through the soft drinks and assorted bottled waters on the shelves.

The Hard Stuff—Beer and Wine

By now, everyone has heard about the research linking moderate consumption of red wine with greater longevity. So, is red wine recommended, or is *any* alcoholic beverage recommended?

For a multitude of reasons, alcohol consumption is not encouraged; if alcohol is consumed, it should be kept to a limit of two ounces per day. That's the equivalent of one beer, one glass of wine, or about one mixed drink.

What about the research on red wine? After closer examination of the studies, scientists realized that the nondrinkers, who died earlier than the drinkers, were more sick to begin with. Due to illness, they were more likely not to drink. Furthermore, some speculate that the social aspects associated with drinking may actually be what is most beneficial—not the alcohol. The social support that comes from enjoying a leisurely dinner or taking a break with family and friends, for instance, may be one key to better health.

The Soft Stuff—(Soda, Pop, Soft Drinks . . .)

You won't find any great choices here either. Many sodas contain caffeine, which we discussed in Aisle 4. Even those that don't are not much better for you. Soft drinks contain large amounts of phosphorus, which increases the body's loss of calcium. Soft drinks are empty-calorie foods, too—not much nutrition in exchange for the calories they provide.

Choose diet drinks, you might suggest. As I discussed in Aisle 7, I do not recommend using aspartame and saccharin, which are used to sweeten diet drinks. But even if you opt to use these sweeteners, there are enough other reasons not to consume soft

---

Talking shopping with Robert Pritikin, director of the Pritikin Longevity Centers, Santa Monica, California:

At night, I sometimes like something hot, and soup is a comfort food. I like our own Pritikin soups. I also like Only a Pinch products—they're stewlike and low in fat, and they are surprisingly low in sodium, too. I like to pick up fresh carrot juice—not because it's healthy or for any other reason other than I just enjoy it. I also like Cutter's nonalcoholic beer. I drink one, ice cold, after my morning workout at the gym. There's something so sinful about drinking a nonalcoholic beer at 10:00 A.M. after a workout! I've served it to some of my beer-drinking friends, and they can't tell the difference between it and regular beer.

drinks more than occasionally. In addition to the caffeine, sugar or sugar substitutes, and the phosphorus, many soft drinks also contain artificial colorings and flavorings.

Here Are the Winners

Beer, wine, and soft drinks may be out, but there are still lots of items on this aisle that are definitely in. For starters, there is a huge assortment of mineral waters available, and many are flavored with natural fruit flavors. Some have no sugar added, and these are great choices. Others have a little sugar—sometimes in the form of fructose—added. These are generally fine, too. They're also sodium and caffeine free.

You'll see tonic water and club soda on this aisle, as well. One bottle of tonic water had an ingredient listing that included water, sugar and/or corn syrup, citric acid, natural and artificial flavors, quinine, and propylene glycol. Another was sweetened with sodium saccharin, so it also contained eighty-five milligrams of sodium per cup.

Quinine is a drug used to treat malaria, and it has not been adequately tested. Stay away from beverages such as the tonic water just described, and as always, stick with foods and beverages with as few chemical additives as possible. Club soda is water with sodium bicarbonate added for fizz. Since it contains a sodium compound, you'll see some sodium listed on the nutrition label—seventy milligrams per cup is typical. One bottle of club soda also contained sodium chloride (salt) and potassium sulfate. Club soda with some sodium bicarbonate is fine. You'll also find seltzer water on this aisle. Seltzer water is mineral water with carbon dioxide added. It's fine.

> **GREAT CHOICES**
> - plain or flavored unsweetened mineral waters
> - lightly sweetened, flavored mineral waters
> - club soda

As mentioned previously, a great way to use mineral water or club soda is to mix it with fruit juice. Some of the sweeter bottled fruit drinks can also be improved by diluting them with mineral water or club soda. You still get the good flavor of the drink, but in effect, you cut back on the sugar and dilute any other objectionable additives that it might contain.

> **Talking shopping with Andy Jacobs Jr., U.S. representative, Indianapolis, Indiana:**
>
> Here is what you would find in my shopping cart at the grocery—when I shop just for myself and not my family. There would be some corn-oil light margarine. There would be a dozen Red Delicious apples. There would be a loaf of fat-free wheat bread. There would be ten bottles of Sundance [kiwi-lime fruit juice with carbonated water], and there would be some frozen veggie patties along with some mild cheddar soy cheese. There would also be some natural peanut butter and some [Simply Fruit] jelly. And of course, there would be a pretty heavy load of broccoli. I look for low or no fat or "light" in any food I purchase.

Aisle 9

Popcorn
Snacks
Snack Crackers

Aisle 10

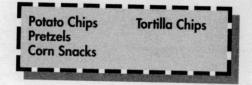

Potato Chips   Tortilla Chips
Pretzels
Corn Snacks

Two aisles of munchies? Remember, this is a tour of a real-life supermarket. And it takes two aisles to hold all of the chips and snacks on the market. Is there a place in a healthy diet for items such as these? We'll take a look at all of the items listed here, and more.

Popcorn

Virtually everyone must be popping their popcorn the modern way, because almost all of the shelf space allotted to popcorn goes to the microwave variety. A meager few bottles and bags of kernels occupy the bottom shelf, an indication that hardly anyone makes popcorn the old-fashioned way anymore.

Microwave popcorn can be a great choice—but you have to make sure that you buy a variety that is truly low in fat. They're out there, but be sure to read labels carefully. If you simply go by the name of the popcorn ("butter light," "all natural, light," or "reduced fat," for instance), you may not be making the best choice. Look at the nutrition information provided on the box and choose the brand that is the lowest in fat. There are choices that contain one, two, or three grams of fat in a three-cup serving—these are the best. Some companies also make salt-free varieties, but it may be harder to find salt-free choices that are also low in fat.

Compare these three microwave popcorn options:

all-natural, light popcorn

Ingredients: popping corn, partially hydrogenated soybean oil, salt, natural flavors	Serving size: 3.5 C
	Calories: 110
	Protein (gm): 2
	Total fat (gm): 3
	Carbohydrate (gm): 20
	Cholesterol (mg): 0
	Sodium (mg): 90
	Calcium (% DV): 0
	Iron (% DV): 4

> **HELPFUL HINT**
>
> To add flavor to popcorn without adding extra fat, spritz hot popcorn with water from a plant mister, then sprinkle with chili powder, cayenne pepper, an all-natural butter-flavored substitute, or another favorite spice or herb mixture. Then toss.

reduced-fat popcorn

Ingredients: popping corn, partially hydrogenated soybean oil, salt, natural flavors, color added, artificial flavors, citric acid and TBHQ added to preserve freshness, butter (sweet cream, water, milk solids)	Serving size: 3 C
	Calories: 45
	Protein (gm): 1
	Total fat (gm): 1
	Carbohydrate (gm): 10
	Cholesterol (mg): 0
	Sodium (mg): 125
	Calcium (% USRDA): 0
	Iron (% USRDA): 2

regular butter-flavored popcorn

Ingredients: popping corn, partially hydrogenated soybean oil, salt, natural flavors, artificial flavors, color added, citric acid and TBHQ added to preserve freshness, butter	Serving size: 3 C
	Calories: 90
	Protein (gm): 1
	Total fat (gm): 7
	Carbohydrate (gm): 10
	Cholesterol (mg): 0
	Sodium (mg): 120
	Calcium (% USRDA): 0
	Iron (% USRDA): 2

The first two choices are good ones in terms of fat content. Either would be acceptable. However, look at the nice, short ingredient list for the first popcorn listed—the all natural, light variety. It contains a little more fat, but it is free of the undesirable additives contained in the reduced-fat option. At a little less than one gram of fat per cup and three grams of fat in a three-and-a-half-cup serving, this would be my personal choice. I would also look around to make sure there wasn't another option that was the lowest in fat and still free of unwanted additives. The last example, the regular, butter-flavored popcorn, is just too high in fat. Skip over a fatty choice like this one.

For those who still prefer to make popcorn in a pan on top of the stove or in a popcorn maker, the key is to use as little oil as possible. With the stove-top method, try to coat the bottom of the pan with just the amount of oil needed to keep the kernels from sticking to the bottom of the pan and burning. Keep the kernels moving by continuously shaking the pan until the popping stops. If you prefer to use a popcorn maker, use a hot-air popper that requires no oil to be added. It'll make fat-free popcorn, and that's the best choice of all.

Snacks and Such

Those little plastic trays containing crackers and spreadable cheese in separate compartments, small packs of cracker sandwiches filled with cream cheese, cheddar cheese, and similar fillings, and a host of other snacks—from beef jerky to novelty chip, nut, and cracker-type mixes—take up a large portion of this aisle. These are generally high in fat and are loaded with salt and undesirable additives. Just keep on walking.

Snack Crackers

See Aisle 1 and Aisle 5 for more information about crackers. As previously mentioned, there are some low-fat and fat-free options located here and on Aisles 1 and 5. Most of these (with the exception of those on Aisle 1) are not whole grain. Whole grain choices would be the best, but the white-flour type are acceptable when whole grain crackers are not readily available.

Remember to watch portion sizes carefully when you evaluate the fat content of a particular brand of cracker. One cracker I found on this aisle had a nice, short ingredient list of whole wheat flour, partially hydrogenated soybean oil, and salt. One serving was listed on the label as three crackers (small ones), which contained two grams of fat. A realistic handful of these might be ten or twelve crackers. Some people could no doubt eat half the box at one sitting (especially if they were planted in front of the television). At two grams of fat per serving, a more typical serving could easily amount to eight or ten grams of fat or more. Too much.

Assorted Chips and Pretzels

I had hoped to find fat-free chips made by natural foods companies on Aisle 10, but I didn't see any here. What I saw were regular potato chips at nine grams of fat for fifteen chips, corn chips and cheese puffs that were equally high in fat, and even a "natural" bag of popped popcorn that had ten grams of fat in a meager one-and-three-quarters-cup serving.

The label on another bag of popcorn said "air popped." I thought it would be low in fat but was flabbergasted to read on the nutrition label that one ounce contained nine grams of fat. This was an eight-ounce bag, which many people could eat in one sitting. If they did, they'd have downed seventy-two grams of fat!

One new product on the market—apple chips—contains four grams of fat in an ounce (about a small handful). These chips are heavier than potato chips, so one ounce doesn't look like as much as one ounce of potato chips. Many people could easily eat the whole three-ounce bag of apple chips in one sitting, and if they did, they'd get a hefty dose of fat: twelve grams. Once again, always consider how realistic the serving size is before you dig in. This product is probably too high in fat for most people, considering how many chips most people would eat.

The savior of this aisle is the pretzel. One brand of sourdough, Bavarian-style pretzel is fat free. The ingredient list—enriched flour, salt, corn syrup, sodium bicarbonate, and yeast—is pretty benign. No whole grain flour here, but for a snack, this product is fine. Other pretzels contain only a couple of grams of fat in a generous serving, and these, too, are great choices.

Bean Dips and Cheese Dips

If you decide to buy low-fat chips and need a dip to go with them, your best bet is to go with any of the many salsas that are found all over the store—on Aisles 1, 2 (page 109), and 4 (page 151), as well as on this aisle. Cheese dips are loaded with fat. If you happen to find one that is fat free (since new products are arriving on the shelves every day), read the list of ingredients carefully to be sure that the product meets your standards in terms of other additives.

As for the bean dips . . . One dip that I found on this aisle contained two grams of fat in a serving. The serving size was small, however: one ounce, which is about two tablespoons—maybe enough for four tortilla-chip scoops. A much better product was actually spotted elsewhere in the store on a freestanding display. This one—from Guiltless

---

GREAT CHOICES

- low-fat or nonfat popcorn
- pretzels
- salsa
- fat-free bean dip

Gourmet—was a fat-free bottled dip. The ingredient list was exemplary: water, black beans, bell peppers, vinegar, sea salt, jalapeño peppers, and spices. (Guiltless Gourmet makes many good fat-free products, including oil-free tortilla chips and other dips.)

Talking shopping with Robert Pritikin, director of the Pritikin Longevity Centers, Santa Monica, California:

There's been a revolution for me at the supermarket. When I was a little kid, there were hardly any snack foods that I could eat growing up in my family. Now I can buy baked potato chips at the supermarket here in L.A. that are fat free. Even five years ago, my wife and I used to slice fresh potatoes and bake them on a nonstick pan to make our own fat-free potato chips. Now I can buy them in the supermarket.

Aisle 11

Frozen Foods

This is the place where grocery carts come to a screeching halt! There has been an explosion of new arrivals on the frozen foods scene—the number and variety of items here is mind-boggling. There are lots of great choices, but we'll need to weed them out from all of the high-fat, low-fiber distracters that occupy this frozen foods jungle.

An Overview—What You Don't Buy Can't Hurt You

My first impression when I took a fast scan of the frozen food cases? *Wow.* Amazement. Almost any meal a person could want can be purchased ready-made—and it can be cooked in minutes using a microwave oven. Eggplant Parmesan, frozen burritos, puff-pastry hors d'oeuvres, chicken pot pie, chicken nuggets, egg rolls, full dinners—you name it, you can probably buy it frozen. Very convenient.

If you are like many people, you are probably well acquainted with all that this aisle has to offer. You may not find it so amazing (and you may wonder what planet I've been living on). But perusing this aisle was an eye-opener for me because my practice is to skip it altogether. My diet and lifestyle have simply evolved that way. I probably haven't bought more than a few frozen dinners in several years. Every so often, I may glide by the frozen foods section to grab a bag or two of vegetables. It's not as if I have to strain to keep myself away from all these convenience foods. Using them is just not a part of my lifestyle. It doesn't have to be part of yours, either.

By foregoing the products on this aisle, I don't give up convenience, though. As I've mentioned earlier, convenience foods for me are a sweet potato that I can bake in my microwave oven and a salad of mixed prewashed salad greens topped with garbanzo beans from a can and a splash of balsamic vinegar. I spend about ten minutes fixing a meal—no longer than it takes to fix some microwave dinners.

What I learned in studying this frozen foods section, though, both confirmed what I had suspected and surprised me by what I had underestimated: namely, that a huge percentage of the foods on this aisle are overprocessed, high in fat, and low in fiber (and, some would add, enveloped in excessive packaging materials, adding to our environmental problems). On the other hand, mixed in with the junk, there are a growing number of wholesome, low-fat foods that are great choices, nutritionally speaking.

Frozen Entrées and Dinners

Have you seen the bags and boxes of frozen ravioli, manicotti, eggplant Parmesan, and other Italian entrées—the ones with no nutrition information on the labels? Read the ingredient lists; most of these are made with high-fat cheeses and eggs. Skip them. Ditto for any of the other frozen entrées or dinners that come without nutrition information— especially the ones that include cheese and cream sauces, such as macaroni and cheese or fettuccine-and-vegetable Alfredo. They're probably loaded with fat, and it's likely that you could find a low-fat alternative elsewhere in this section.

When you evaluate the frozen entrées and meals that are available, take a look at the nutrition information first. Is this item low in fat? "Low in fat" is a relative term, of course. If the product contains twenty grams of fat in a serving, and this is the only fat you consume all day long, then maybe it's okay. Otherwise, aim for foods that are as low in fat as possible—five grams of fat per serving or less is a pretty good goal, if the serving size is realistic.

The issue of serving size is definitely something to consider carefully. I purchased and sampled about twenty different products from this section in preparing this book, and only once did I find the portion size of the product to be realistic. Now, I am not a large person with a colossal appetite. But in all cases where frozen entrées were listed as "two servings" per box, I ate both at one meal.

So, as always, take into consideration the number of servings you are likely to eat at one time. A five-gram-per-serving food may sound good at first, but if the reality is that you would eat two or three servings at one sitting, then you may not want to buy that product.

Judging a frozen entrée or dinner by its fat content is a good first step. But especially with items that are likely to be a major part of your meal (or the whole meal), you should consider more than just the fat content. Read the list of ingredients and take a moment to think about the ones that make

┌───┐
│ **HELPFUL HINT**   │
│   │
│ When frozen entrées or dinners are not enough to │
│ satisfy you, round out the meal with such foods as a │
│ salad made from fresh ingredients and a nonfat dressing, │
│ a slice of whole grain bread, and/or some fresh fruit. │
└───┘

up the bulk of the dish or meal. Are they fiber rich, such as vegetables or whole grains? Or is the product made primarily of refined grains, such as white-flour pasta, and ingredients of animal origin, such as cheese or eggs? Choose products that are made of mostly vegetables, grains, beans, or fruit. These are the foods that are the most nutritious and contain the least amount of fat (as well as no cholesterol).

Another quick way to help you judge the relative merits of one product over another is to look at the nutrition label and check the cholesterol content. This can be a clue to how much or how few animal-product ingredients the food contains. A food that contains no cholesterol, or very little, is more likely to be fiber rich and healthy. On the other hand, foods that contain higher cholesterol contents are more likely to be heavy on the cheese and/or eggs and to be low in fiber.

Take a look at the following two products:

zucchini lasagna

Ingredients: tomato sauce, green zucchini, lasagna pasta sheets, tomatoes in tomato puree, diced tomatoes in juice, cottage cheese, part-skim mozzarella cheese, modified food starch, onions, sugar, Parmesan cheese, spices, granulated garlic, granulated onions, xanthan gum, salt, citric acid	Serving size: 1 meal Calories: 330 Total fat (gm): 1.5 Protein (gm): 20 Carbohydrate (gm): 58 Cholesterol (mg): 10 Dietary fiber (gm): 11 Sodium (mg): 310 Calcium (% DV): 20 Iron (% DV): 15

fettuccine primavera with tortellini in light cream sauce

Ingredients: in pasta pouch—
cooked fettuccine noodles,
cooked tortellini (wheat flour,
water, eggs, dough mix, ricotta
cheese, Parmesan cheese,
Romano cheese, fontina cheese,
parsley, salt, spices), margarine;
in sauce pouch—heavy whipping
cream, water, cream cheese,
margarine, half-and-half cream,
Parmesan cheese, Romano
cheese, sherry wine, stabilizer,
salt, flavoring, spice, paprika; in
vegetable pouch—broccoli,
cauliflower, red peppers, carrots

No nutrition information given

How would you evaluate the first product? Only
a gram and a half of fat per serving—not bad. And
no wonder—the first ingredients are tomato sauce,
zucchini, pasta, and tomatoes. Cheese comes in a
little further down the list and accounts for the ten
grams of cholesterol in a serving. Considering that
this dish is so low in fat and that its primary ingredients are of plant origin, this is a great choice. This
product also contains eleven grams of fiber, which
is a reasonably good amount. Add a slice of whole
grain bread, a small mixed green salad, and a piece
of fresh fruit for dessert, and this would be a good
meal.

Now take a look at the second product. The first
red flag should be the lack of nutrition information.
Then any suspicions that this product might be high
in fat are reasonably well validated after reading the
list of ingredients: egg and cheese pasta with margarine added, and a fatty cream sauce. The vegetables are the only healthy ingredient here. This
product is an example of one to avoid.

Some frozen entrées and meals are very low in

fat, but they contain meat or seafood. Are these good choices?

In some cases, where meat or seafood play a very minor role in a dish, that dish can still be high in fiber, low in fat and cholesterol, and quite nutritious, despite the fact that meat is present. So such dishes could fit into a healthy diet.

My recommendation, however, is generally to steer away from meat-containing dishes, no matter how little meat the dish contains. The meat that is contained in the dish still displaces plant matter that would probably be more beneficial. But much more significant is simply the matter that by getting into the habit of eating a low-fat, vegetarian diet, you will be establishing an eating pattern that is broadly health supporting, and you will be less likely to slide back into old patterns of eating.

Here's another situation to consider: I spotted a child's frozen meal that appeared to be low in fat and reasonably high in fiber, even though the entrée and dessert were foods that are typically high in fat and low in fiber. The meal consisted of macaroni and cheese, stewed apples, corn, and a fudge brownie. The whole meal contained 310 calories, seven grams of fat, fifteen milligrams of cholesterol, and six grams of dietary fiber. A good choice?

First, keep in mind that this meal would be higher in fat if a pat of margarine was added to the corn (which many people would do out of habit). So, as always, added fat is not recommended. Barring that, this meal *is* relatively low in fat. Since you know that each gram of fat contains nine calories (the new food labels give this information), and there are seven grams of fat in this meal, then about sixty-three calories in this meal come from fat. Sixty-three calories out of a total of 310 calories in the entire meal equals about 20 percent of calories from fat— not bad for a child's diet. The cholesterol comes from the egg and dairy products used in this meal. The fiber comes primarily from the corn and apples.

Out of curiosity, I actually purchased this product and tried it. Once heated, I saw that the serving size of the macaroni and cheese was very small—not much more than each of the other individual components of the meal. Since macaroni and cheese is typically very high in fat, keeping the serving small helped to keep the fat content in check. The brownie was also tiny. A meal such as this one can be fine for a child.

In short, there are lots of frozen entrées and meals available. To evaluate them, consider not only the fat content (as low as possible—aim for five grams of fat per serving or less), but also consider the cholesterol content (preferably five milligrams per serving or less), the ingredients (whole grains? vegetables? beans?), and the portion size for one serving (is it realistic?).

Talking shopping with Jennifer Raymond, nutritionist and author of *The Peaceful Palate* cookbook, Calistoga, California:

If you ran into me at the supermarket, you'd probably find me wandering in blissful contentment down the produce aisles. I find this area absolutely inspirational with its abundance of wonderful colors, shapes, and smells. There's nothing like fruits and vegetables, piled high, to inspire me to cook.

One of the things you'll *never* see me doing at the supermarket is using coupons. Coupons don't really save you money because they encourage you to buy high-priced, highly processed foods. Think about it— you never see coupons for carrots, or broccoli, or dried beans.

I spend about twenty dollars a week on food for the two of us. The way I keep my food bill low is by sticking to grains, beans, vegetables, and fruits in the most unprocessed forms I can find.

Sweet Potatoes and Potatoes

The ingredient list on a box of microwavable candied sweet potatoes reads: sweet potatoes, candied sauce mix, dextrose, brown sugar, corn syrup solids, cellulose gum, and caramel. Not bad. This product, as packaged, is fat and cholesterol free.

But the directions on the box say to add two tablespoons of margarine. That's thirty grams of fat! The label also says that this box makes three servings. Hmmm. I ate the *whole box* plus another frozen entrée for dinner one night. Together, both products made a reasonable-sized meal.

The sweet potatoes as I fixed them, however, were fat free. Although the instructions called for margarine to be added, I added a couple of tablespoons of lime juice instead. The dish was delicious. Prepared my way, this product is a great choice.

Here's another example:

stuffed potatoes

Ingredients: fresh potatoes, water, margarine, natural cheddar cheese, salt, nonfat dry milk, dehydrated pasteurized processed cheddar cheese, onions, pepper, paprika	Serving size: 1 half potato
	Calories: 130
	Total fat (gm): 4
	Protein (gm): 4
	Carbohydrate (gm): 20
	Cholesterol (mg): 0
	Dietary fiber (gm): 4
	Sodium (mg): 310
	Calcium (% USRDA): 6
	Iron (% USRDA): 4

This product is basically mashed potatoes with cheddar cheese and margarine added. The fat content is moderate. But consider this: it takes nine minutes to heat this product in a microwave oven. It would take less time than that to heat a large fresh potato in the microwave—and there's no fat at all in *that*. This one's a no-brainer—take the fresh potato and add some salsa, ketchup, nonfat yogurt, and pepper, or nonfat, butter-flavored sprinkles instead.

What about other potato products? Items such as tater tots, hash browns, french fries, and potato wedges universally have fat added to them. One small box of microwavable french fries that I picked up contained twenty grams of fat! Some packages also call for extra fat to be added—but of course, you can always leave that out.

When it comes to potatoes, you are almost always better off using fresh potatoes and fixing them yourself. Go back and look at the recipe for potato wedges in Aisle 2. We're not talking any big sacrifice in preparation time here, either. It takes only seconds to wash a whole potato and set it in the microwave oven to cook, and it takes only a few minutes to make your own potato wedges. And fresh potatoes take no longer to cook—or even less time—than do frozen, prepared potatoes. As we discussed earlier in this book, there are lots of nonfat potato toppings, too, such as nonfat yogurt, salsa, and ketchup, for starters.

Frozen Fruit and Pies

PASS (OVER) THE PIE, PLEASE

Most fruit pies and cobblers contain eight to ten grams of fat per slice—and those are *small* slices, by my standards. The problem lies mostly in the crust, which is full of shortening. You'll also see frozen, ready-made pie crusts here, along with fat- and cholesterol-rich quiches and quiche mixes that are ready to be poured into fatty pie shells. Bypass all of these.

One ready-made cheesecake-like pie topped

with sliced peaches advertised that "you can enjoy an incredibly luscious dessert and maintain your healthy lifestyle" and claimed that this product is "perfect for an everyday indulgence." One serving contained five grams of fat and five milligrams of cholesterol. I have a big sweet tooth; I took one home to check it out.

I wasn't surprised to find that inside the package, this pie was smaller than life. According to the package, this pie contained six servings. I couldn't be happy with a slice less than *double* that. No, this was definitely not going to be an everyday indulgence for me.

PIE TOPPERS

While we're at it, we might as well take a look at nondairy whipped toppings, too. Are any of them low fat? Yes. In fact, most light whipped toppings contain less than one gram of fat per tablespoon and no cholesterol. But take a look at the ingredient list of a typical brand:

light whipped topping

Ingredients: skim milk, corn syrup, partially hydrogenated palm kernel oil, hydrogenated coconut oil, sugar, sodium caseinate, water, dextrose, artificial flavor, polysorbate 60, sorbitan mono-stearate, xanthan gum, guar gum, disodium phosphate, colored with turmeric and annatto extracts

The polysorbate 60 is an emulsifier that is considered safe. The palm and coconut oils are saturated fats, but they are present in such small quantities that they do not add up to much in the amount of this product that is commonly used. Aside from the artificial flavor, the other ingredients are fairly benign. This product contains no fiber or ingredients that are particularly healthful—it's a confection, a nutritionally empty so-called food. Used in small amounts—a dollop on a bowl of berries, for instance—it's fine, though.

Talking shopping with Patti Breitman, literary agent, Fairfax, California:

I'm afraid that supermarkets in California may not be typical nationwide. I buy jars of papaya already cut up. Also, I buy Popsters—a low-fat baked potato chip that's delicious. I buy salsa sometimes, and lots of oranges to squeeze my own juice. . . .

Here's what Stan and I do when we're on the road: we buy ripe red peppers and eat them like apples. They're delicious!

FROZEN FRUITS

When peaches are out of season, you can always buy them frozen. In fact, there is a wide variety of frozen fruit available, and they're all great choices. You can use frozen fruit to make your own low-fat, fruity desserts, fruit smoothie drinks, fruit salads, and so on. Choose from cherries, blueberries, pineapple, red raspberries, mixed fruit, peaches, strawberries, and berry mixes.

Frozen Veggies

Nutritionally, frozen vegetables are very similar to fresh. For convenience's sake, many people find it easier to keep frozen vegetables at home, rather than large quantities of fresh vegetables that might spoil before they can be eaten.

There is a huge variety of frozen vegetables available, and most are excellent choices. You'll find all the usuals—broccoli, spinach, corn, various mixtures of different vegetables, and some that are more seasonal or otherwise less likely to be found fresh, such as okra and certain types of beans.

Some frozen veggies, such as certain stir-fry mixtures, come with a seasoning packet that contains a variety of spices. Other products come with butter sauce, cream sauce, or cheese sauce. Sauces like these can add three grams of fat per half-cup serving, so avoid them. A half cup isn't a large serving at all, and on a healthy, low-fat diet, it's more likely that you will be eating hefty, one-cup servings of vegetables. In that case, an extra six grams of fat in a one-cup serving of vegetables is a significant amount. Steer clear of added fats whenever possible.

Frozen Meats

See The Meat Counter.

Talking shopping with Spice Williams, stuntwoman and actress (Spice costarred as Vixis in *Star Trek V*), Van Nuys, California:

If you met me at the checkout line at the grocery store, what would I be likely to have in my cart? Pineapple, mangoes, papayas, kiwis, strawberries, grapes, apricots, plums, oranges, bananas, and if in season, persimmons and pomegranates. Gourmet lettuce, tomatoes, mushrooms, green and red peppers, red cabbage, red onions, sunflower seed sprouts, broccoli, cauliflower, brussels sprouts, Swiss chard, mustard greens, string beans, eggplant, crookneck squash, and zucchini. Potatoes, yams, short sweet brown rice, barley, millet, corn tortillas (made without lard or oil), pasta, whole grain bread . . .

I'm happy to say that some Los Angeles supermarkets are slowly becoming health-conscious. I now can actually go to a local Pavilions Market and buy my soy milk, veggie burgers, tofu hot dogs, and soy or almond cheese.

Frozen Breads, Rolls, Croissants

There are some great choices to be found here, along with some dogs (literally). Meat- and cheese-filled croissants, for instance, pocket sandwiches made with meat and cheese, steak sandwiches, corn dogs, and chili dogs are some prime examples of fat-laden finds in this section. A few other convenience items are tucked in here as well, such as microwavable hamburgers, breaded fish sandwiches, and bagel pizzas topped with meat and cheese. These aren't so tough to spot for what they are—by now you can probably easily identify some of the most likely fatty suspects. Nutrition labels will confirm your suspicions in most cases, and in the absence of nutrition information, a quick look at the ingredients list is usually all that is necessary to rule out some of these products.

What about the breads themselves? Frozen pita bread, or pita pockets, are located here, and they're a great choice (especially the whole wheat variety). Pita bread can be used to make sandwiches, and when cut into wedges and lightly toasted in the oven, they make great scoops for salsa or nonfat black bean dip.

Read the nutrition label before you buy a loaf of frozen garlic bread. These breads are usually slathered with butter and are high in fat. A quick survey of the ones I found in my supermarket found that they averaged eight to nine grams of fat per slice!

Instead of buying ready-made garlic bread, you would be much better off simply buying a loaf of Italian bread, which is virtually fat free, cutting the loaf in half lengthwise, then spraying a thin coat of vegetable oil spray on the insides of each half. Smear each half with some minced garlic, and heat the bread in the oven until it is warm.

Some of the ready-to-bake dinner rolls that you will find in this section are relatively low in fat—one gram of fat per roll. Buy whole grain varieties if you can find them. Ignore instructions to brush the rolls with butter before baking. A light squirt of vegetable oil spray will add less fat—or skip the oil altogether.

Another great choice: bagels. You can't go wrong. All bagels are low fat. The only type you might want to avoid is egg bagels, which have ten milligrams of cholesterol per bagel—not much, but then why bother when there are so many other varieties to choose from?

Frozen Pizzas

Envision a pizza the size of a dinner plate. Now envision a slice the size of about one quarter of that. A typically sized piece of pizza, right? Well, one slice of a standard frozen meat-and-cheese pizza purchased in the supermarket contains about sixteen grams of fat. Wow. Another three-cheese variety that I saw contained fifteen grams of fat in half of an eight-inch pizza. Many people would eat the whole pizza—not an unreasonable amount, unless they stop to figure that they're getting a whopping thirty grams of fat along with it.

I looked at many varieties of frozen pizza. One was billed as "light" and contained ten grams of fat per slice. Some that were advertised as being low in fat and cholesterol contained four grams of fat per serving, but the slices were very small. With four servings per box, most people would eat all four slices at a meal, therefore downing sixteen grams of fat from pizza. See The Deli/Bakery and Aisle 4 for more information about pizza. Also, Aisle 12 contains some ready-made pizza crusts that can be topped with seasoned tomato sauce and your favorite veggies—a quick, healthy, low-fat pizza option that takes only minutes to prepare. So, skip the frozen varieties unless you find one that is truly low in fat.

Danish, Muffins, and Cakes

If you succeeded in avoiding the cakes and pastries in the deli/bakery area of the store, then you have a second chance to dodge them here on Aisle 11. Here you will find frozen cheesecakes, cakes of all varieties, cinnamon rolls, pastries, and more—most of which are quite high in fat. The fruit dumplings wrapped in puff pastry, for instance, weigh in at eleven grams of fat apiece. Some toaster pastries contain seven grams of fat each.

If you scout around, however, you will find a few items worth considering. A few companies make muffins that contain only two grams of fat each, and these are fine. Here's an example:

raisin bran muffins

Ingredients: unbleached wheat flour, unrefined cane sugar, egg whites, water, raisins, wheat bran, unsulphured molasses, partially hydrogenated soybean oil, baking soda, nonfat dry milk, modified food starch, polyglycerol esters, salt, guar gum	Serving size: 1 muffin
	Calories: 140
	Protein (gm): 4
	Total fat (gm): 2
	Carbohydrate (gm): 30
	Cholesterol (mg): 0
	Dietary fiber (gm): 4
	Sodium (mg): 260
	Calcium (% USRDA): 4
	Iron (% USRDA): 10

As always, if you can find muffins that are low in fat *and* made with whole grain flour, that's even better.

You will also see here some specialty desserts made by manufacturers that are associated with weight-control programs and lower-calorie products. Read the labels carefully, and remember to pay close attention to portion size. One item, for instance—a chocolate chip cookie dough sundae—comes two to a box. Each sundae contains four grams of fat. The servings are *small*. Realistically, many people would find it hard not to eat both at one time. In that case, you would need to count this item as containing eight grams of fat and decide accordingly whether it fits into your diet or not.

Another consideration: once you open the packages, it can be surprising how little food is actually contained inside. As I noted earlier, from an environmental standpoint, some people have a problem with the amount of excess packaging that is used for many frozen products. They're expensive, too, considering how much food (very little) the packages contain.

We buy these products for their convenience, but that advantage should be weighed against cost and other factors, such as the environmental impact. In some cases, you might even decide that the convenience of some of these products is overrated. For example, it is a snap to make a dozen muffins at home, and they can easily be frozen, taken out as needed, and reheated. Making your own is much cheaper than buying them ready-made, and you have the added advantage of being able to control the ingredients—to use whole grain flour and less fat, for instance. Just food for thought . . .

Meat Substitutes—Veggie Burgers and Dogs

Some exciting developments have been taking place in the area of meat substitutes—products that are made of vegetables, beans, and/or grains but that resemble their meat counterparts in appearance, taste, and/or function. There are now a large number of products from which to choose. Examples include bacon, hot dogs, breakfast sausages

> **HELPFUL HINT**
>
> If you are bringing veggie burger patties or veggie dogs to a cookout or picnic, be sure to bring more than you actually need. You can count on everyone else wanting to try yours; many people will prefer yours when they see how good they taste and hear how low in fat and cholesterol they are compared to their meat counterparts. Bring extras so that you'll have enough for yourself.

and links, burger patties, and others—many made with soy, rice, or vegetables and containing no meat, no cholesterol, and no saturated fat.

Most of these products are far lower in total fat than their real-meat counterparts. Some of the veggie burger patties, for instance, are fat free, and some contain only a few grams of fat per patty. Compared to the typical fifteen grams of fat or more in a regular hamburger patty, this is a huge improvement.

These products qualify as processed foods, though, so they contain very little fiber and can be high in sodium. Nevertheless, they may still have their place under certain circumstances. They're certainly convenient. Kids love them. They can be especially convenient on summer holidays, picnics, or occasions when a group of people wants to cook foods outside on a grill. When everyone else is eating fatty steaks, hamburgers, and hot dogs, any of these meatless veggie burgers or dogs can be used in place of the real thing.

There are several different manufacturers making these products, and each brand has several varieties from which to choose. Experiment to find your favorites, but stick with the ones that are lowest in fat.

Frozen Breakfast Foods

We've covered frozen dinners, but there is also a world of frozen breakfast meals—microwave scrambled-egg-and-bacon breakfasts, breakfast burritos, and French toast sticks.

Guess what they all have in common? That's right: fat, and lots of it. There's less variety here than there is in the frozen-dinner arena, and not much to recommend. Some of these products can be tremendously high in fat, in fact. One box of French toast sticks, for instance, has fourteen grams of fat in one wimpy serving—and that's without the margarine that many people would add at the table. Eggs figure prominently in many of these products, boosting the saturated fat and cholesterol content sky-high.

EGG SUBSTITUTES

What about egg substitutes?

The frozen egg substitute found on this aisle is basically egg whites, often with yellow food coloring and a few other minor ingredients added. It comes frozen in small milk-container-like boxes and can be thawed in the refrigerator or microwave oven before being used in the place of liquid whole eggs. A quarter of a cup of egg substitute usually substitutes for one whole egg in recipes calling for eggs.

As I've noted throughout this book, there are many other foods that can be used in place of eggs in recipes. Again, eggs serve as a binder in the recipes that call for them, but bananas, tofu, and commercial powdered egg substitutes (arrowroot starch) can also be used in their place (see pages 56 and 81).

The frozen egg-substitute product is cholesterol free, since it does not contain egg yolks, and it is typically low in fat. When used in small amounts—

the equivalent of one or two eggs in a loaf of bread or casserole, for instance—it can be fine. I encourage people to try the alternatives I've listed, however, since becoming familiar with them helps to break the reliance on and the habit of using animal products in the diet. Using larger amounts of egg substitute—for omelets, scrambled eggs, and so on—is not recommended, for much the same reasons that were given earlier regarding meats (see The Meat Counter). Meals should be plant-based, with animal products making up no more than a minor part of the meal. If you do want an omelet or scrambled eggs on occasion, then limit the egg substitute to the equivalent of two whole eggs. You might want to add chopped vegetables, such as spinach, onions, tomatoes, and mushrooms to increase the portion size of the dish without resorting to more egg product.

HELPFUL HINTS

- If you prefer not to use commercial egg substitutes, which frequently contain yellow food coloring, you can use two egg whites to replace one whole egg in most recipes.
- An alternative to scrambled eggs is scrambled tofu. Use reduced-fat tofu and season with herbs and spices. (At the time of this writing, a variety that contains 1 percent fat and is packaged in aseptic cartons is available in natural foods stores but has not yet made it into the supermarket.) A popular seasoning mix for scrambled tofu can be found in the natural foods brands section (Aisle 1 in my store). Scrambled tofu can be eaten in place of scrambled eggs, and it can be used as a sandwich filling in whole grain pita pockets or on a crusty Kaiser roll. Add some veggies to the dish (as noted earlier for scrambled eggs and omelets).

WAFFLES AND MICROWAVE PANCAKES

Microwave pancakes are available, ready-made and ready to heat, in the frozen foods section. Those I found were made with white flour, and they had four grams of fat in three small pancakes—a little over a gram of fat per pancake. Not bad, unless you add extra margarine to your pancakes at the table. It would be better if they were made with whole grain flour. By making your own pancakes from scratch or from a whole grain mix, you can freeze the extra and reheat them later (just as with the muffins earlier).

The case is much the same with frozen waffles, although some are made, at least in part, with whole grain flour. Here's an example:

whole grain waffles

| Ingredients: whole wheat flour, enriched wheat flour, whey, water, egg whites, unsweetened applesauce, partially hydrogenated soybean oil, baking powder, wheat bran, salt, soy, soy lecithin | Serving size: 1 waffle
Calories: 90
Protein (gm): 3
Total fat (gm): 3
Carbohydrate (gm): 14
Dietary fiber (gm): 2
Sodium (mg): 200
Calcium (% USRDA): 2
Iron (% USRDA): 10 |

This example is moderately high in fat, and if extra fat were added at the table, this could end up being a high-fat meal. But at three grams of fat per waffle, a couple with breakfast could fit in acceptably for some people, especially if the rest of the meal is fat free. Add a little maple syrup or applesauce and some fresh fruit, for instance.

I did see one variety of waffle that was advertised as being low fat. It contained less than one gram of fat per waffle, which was great. However, it was made with all white flour, artificial coloring, and artificial flavor. Once again, the decision to eat a product like this is a personal one; each person has to weigh the importance of a variety of factors. Nutritionally, this product is probably a neutral and would be fine to use on occasion.

Frozen Juices

It's tough to go wrong here. Generally speaking, you can go for it—whether your favorite is pineapple-orange juice, Granny Smith apple juice, pink lemonade, or cranberry-raspberry juice. Many of these frozen concentrates are 100 percent fruit juice, with no sugar added. These are the best choices.

Some would argue that the juices that are 100 percent juice are no different nutritionally than those that are more or less sugar water with a smaller percentage of real juice, which are often enriched with vitamin C. For these people sugar is sugar, whether it comes from fruit (fructose) or table sugar (sucrose). Furthermore, many of the 100 percent juices are made up largely of white grape juice anyway, which has little nutritional value.

Using this logic, we might say that a Twinkie (which is fortified with vitamins and minerals) would be more nutritious than an apple. I reject this line of reasoning. I always hold out for the real thing, and I think that it's a good habit to choose whole foods first, whenever possible—the least refined, the least processed, the most natural products. For the same reason that I recommend choosing 100 percent juices first over other fruit beverages, I also encourage people to choose whole fresh fruit in place of fruit juice when there's a choice.

With that said, there's still no reason that most of the juices sold here can't be enjoyed to some extent. Whether they're "100 percent juice" or not, all of these juices are concentrated sweets, and it's not a bad idea to dilute them as I discussed in Aisle 6, where the bottled juices were located. Pour a

---

GREAT CHOICES

- low-fat frozen entrées and meals (especially those made primarily of vegetables, whole grains, beans, and fruits, about five grams of fat or less per serving, and five milligrams of cholesterol or less)
- fat-free frozen sweet potatoes
- plain frozen fruits: cherries, blueberries, pineapple, red raspberries, mixed fruit, peaches, strawberries, berry mixes
- plain frozen vegetables: broccoli spears, chopped broccoli, leaf or chopped spinach, cauliflower, cut corn, green beans, mixed vegetables, lima beans, green peas, carrots, pea pods, brussels sprouts, stir-fry mixes, okra, beans, turnip and mustard greens, corn on the cob
- frozen pita pocket bread, especially whole wheat
- frozen low-fat dinner rolls, especially whole wheat
- frozen bagels
- low-fat frozen muffins, especially whole grain
- fat-free and very low-fat meat substitute products, such as veggie burger patties and veggie dogs
- frozen egg substitute in small amounts (limit to the equivalent of two whole eggs per day)
- low-fat frozen waffles and pancakes, especially whole grain
- frozen juice concentrates: pineapple-orange, grapefruit, lemonade, Granny Smith apple, cranberry-raspberry, grape, daiquiri mix, limeade, and others (avoid coconut milk)

---

few ounces of juice into a glass, then fill it up the rest of the way with sparkling mineral water or seltzer water.

Some people might enjoy drinking the margarita, daiquiri, and rum drink mixes this way, too. The only exception to these juices is the piña colada mix, which contains coconut milk. Coconut is high in tropical oil, a saturated fat.

Aisle 12

Dairy	Cones
Ice Cream	Bread
Ice	

We're now rounding the bend and coming to the last leg of this supermarket tour. We'll take a look at fresh pastas again and their toppings. Then, we'll check out the dairy case and discuss milk, cheese, yogurt, eggs, and egg substitutes. This is also where you'll find the refrigerated cookie dough, biscuit dough, and more ready-made pizza crusts.

We'll move along and compare breads, buns, and English muffins. Finally, what's one of the top bedtime snack choices in the United States? You guessed it! Appropriately, our tour will finish with the food with which most Americans are likely to finish their day: ice cream (and Popsicles, frozen fruit bars, and so on).

More Fresh Pasta, Fresh Toppings, Fresh Pizza Kits

If you were holding out for fresh, cholesterol-free pasta and didn't find it in the deli/bakery area, there's another chance to find it on Aisle 12, in the refrigerated section. While some cholesterol-free pastas are entirely egg free, the fettuccine and linguine found here contain egg whites, no yolks. It's the yolk that contains the fat and cholesterol in eggs, so this egg-white pasta is cholesterol free and low in fat. This kind of pasta is a good choice.

Now, this pasta is also made with white flour. I didn't see any made with whole wheat at my store. And in case you wondered, the green, spinach pasta (or any other vegetable-colored pasta) is essentially the same nutritionally as the white-flour type. So, as always, a whole grain choice would be optimal. But you can compensate a little by topping one of the other types with a fat-free sauce and/or lots of veggies, making it the foundation for a very healthy meal.

THE TOPPINGS

Your choice of a pasta topping can make or break a meal—nutritionally speaking, that is. We took a look at some pasta toppers on Aisle 4; now we see a few more in the refrigerated section of Aisle 12. The best choices? Look for a tomato-based sauce with little or no oil added—there should be not more than a gram or two of fat in a serving. Alfredo and pesto sauces tend to be loaded with fat from oil or fatty dairy ingredients. Take a look at two examples of fresh sauces found on this aisle:

light alfredo sauce

Ingredients: Skim milk, cream, water, asiago cheese, rice flour, Parmesan cheese, rice starch, salt, spices	Serving size: ½ C
	Calories: 190
	Protein (gm): 8
	Total fat (gm): 13
	Carbohydrate (gm): 10
	Dietary fiber (gm): 0
	Sodium (mg): 560
	Cholesterol (mg): 40
	Calcium (% DV): 25
	Iron (% DV): 0

tomato basil sauce

Ingredients: plum tomatoes, tomatoes, tomato puree, onions, cream, tomato paste, basil, celery, olive oil, salt, garlic, spices	Serving size: ½ C Calories: 70 Protein (gm): 2 Total fat (gm): 3 Carbohydrate (gm): 8 Dietary fiber (gm): 3 Cholesterol (mg): 0 Sodium (mg): 450 Calcium (% DV): 4 Iron (% DV): 6

Looking at the nutritional information for the light Alfredo sauce, you may have done a double take. Thirteen grams of fat in a half cup?! If that's light, I'd hate to see the regular variety. Of course, maybe this is one of those cases where light really means light*er*, as in texture or flavor. I'd turn my nose up on this one and pass it by.

Now, the tomato basil sauce, on the other hand, has a little more going for it. It's made with mostly wholesome ingredients. With just a small amount of cream and added oil, it tips the fat-o-meter at only three grams of fat per serving. This is a much better choice. On the other hand, you could easily fix this sauce yourself at home, mixing all the same ingredients but leaving out the cream and the oil for a fat-free marinara sauce. Make extra and freeze some for later.

MORE PIZZA KITS

You'll find the odd refrigerated pizza crust here—ready for your own toppings. See The Deli/Bakery, Aisle 4, and Aisle 11 for more discussion of pizza. The few kits I found here came equipped with cheese and tomato sauce toppings, and some included pepperoni. The cheese kit—made with part-skim mozzarella cheese—contained eleven grams of fat in one quarter of the pie (one big slice)—too much fat for most of us.

The plain crust, for build-it-yourselfers, did not

Talking shopping with Joyce Goldstein, chef and proprietor of Square One restaurant in San Francisco and food columnist for the *San Francisco Chronicle*, San Francisco, California:

If you met me at checkout land, you'd find the following in my basket: salad greens, greens, any vegetable that is in season (artichokes, asparagus, beans, et cetera), potatoes, garlic, onions, yogurt, bananas, berries, peaches, apricots, pears, etc., dried apricots, bagels, baguettes, tortillas, chilies, assorted cheeses (sorry, I love them too much to give them up), olive oil, vinegars, dried pasta, rice, couscous, lemons, oranges, fresh herbs. . . .

I eat at the restaurant most of the time and occasionally prepare dinner for my family on Sunday, my day off. I like to make fruit and yogurt shakes when I am working at home as they are fast and filling. But pasta and salad are a way of life for me. I do drink a glass of red wine almost every day with my dinner. I don't buy any pre-prepared foods, and I find that as the years go by I eat less food. I get full fast and I rarely eat dessert.

include nutrition information. Remember, if you use one of these crusts, read the label first, then stick to a crust that contains minimal added oil (aim for crusts with two grams of fat per slice or less), and choose one made with whole wheat flour when it's available. Then top it with a fat-free tomato-based sauce and veggies.

The Dairy Case—Cheese, Milk, Butter, and the Rest

FIRST CASE: MILK AND MILK PRODUCTS

Despite the many products that you'll find in this section, my discussion of them is going to be comparatively brief. Once more, the bottom line first: there is no human requirement for milk from a cow. The use of milk and its products in our country is strictly a cultural tradition; there are millions of people around the world who never consume cow's milk and are none the worse for it. Milk does contain vitamins and minerals, but all of these can easily be gotten from other foods.

Milk—especially whole-milk products—is also a concentrated source of nutrients that Americans tend to get in excess, such as fat, cholesterol, and protein. Milk and milk products contain no fiber.

It is still common practice for nutritionists in the United States to recommend that people—especially teens and women—include dairy products in their diets, primarily for the calcium. The nutritional issues surrounding recommendations for calcium in the diet were discussed in Nutrition in a Nutshell and are somewhat complex and fraught with scientific uncertainties. My recommendations for the use of dairy products are summarized in the Helpful Hints that follow.

If you use dairy products, use only nonfat varieties—and limit the use of those, too. What about 1% or 2% milk and yogurt? Too high in fat for most of us. The terminology used to describe the fat content of milk is deceptive and refers to the percent of fat in the product by weight, not by calories. Here's an example that will help explain how this works:

Have you ever seen frozen yogurt advertised as being 96 percent fat free? Sounds pretty good, right? Well, this product contains the same amount of fat as does whole milk. Now, most people understand that whole milk is high in fat. Milk is more than 80 percent water. In whole milk, over 50 percent of the calories come from fat. What if I turned around the description of whole milk, and instead of call-

┌───┐

HELPFUL HINTS

- If you choose to use milk or milk products, use only nonfat varieties and limit intake to one to two servings per day. A serving equals one cup of milk or yogurt or one ounce of nonfat cheese.
- There is no evidence that people who consume a low-fat vegetarian-style diet and who also take in less than the recommended dietary allowance for calcium are at any additional risk for osteoporosis. However, if you and your health care provider are uncomfortable with your calcium intake, then you may want to consider taking a calcium supplement.
- For those who want to take a calcium supplement and who use nonfat dairy products and/or regularly eat plenty of calcium-rich plant products, such as green leafy vegetables, whole grains, and legumes, a supplement offering five hundred milligrams per day should be more than sufficient. Junk-food junkies can take up to two thousand milligrams of calcium per day. (But if you don't eat well, you are likely to be getting insufficient amounts of a lot of nutrients, not just calcium! Eat well. Plan to get the nutrients you need from whole foods, not supplements.)

└───┘

ing it "4 percent fat," I said it was 96 percent fat free? This is what is done when frozen yogurt is described as being "96 percent fat free." Even half the fat—as in 2% milk—is too much fat.

"Okay," you say. "You've put the kibosh on my extra-sharp cheddar cheese—or any cheese worth eating, for that matter—and I can kiss my ice cream and the cream in my coffee good-bye. What about fat-free Parmesan cheese, fat-free cream cheese, and fat-free sour cream?"

I don't see a problem with using tiny amounts of

any of these, unless it is simply that these products can become a crutch for some people and prevent them from making more fundamental changes in their food choices. But used as a condiment—truly in very small amounts—a sprinkle of fat-free Parmesan cheese over your pasta or a smidgen of fat-free cream cheese on a bagel is no big deal.

BUTTER AND MARGARINE

The score here is the same as it was for oils. Don't use added fat—even reduced-fat margarine or butter. Get used to putting honey or jam on your toast and to using maple syrup, minus the blob of butter, on your pancakes. Just think: you can avoid the whole trans-fatty acids controversy, and you don't have to give the issue of saturated versus polyunsaturated a second thought. Not to be flip, but these issues are really not significant when you consider the big picture, and the big picture is that there isn't room for these added fats anyway.

As for the fat-free margarines now available: same story as for the fat-free Parmesan cheese and cream cheese. If you really want to, go ahead. Some people, such as ethical vegetarians, will want to avoid them anyway because they may contain gelatin. They may also contain artificial flavor.

MORE LITTLE EXTRAS—WHIPPED TOPPINGS AND COFFEE CREAMERS

As with other high-fat dairy products, skip the whipped cream. We covered nondairy whipped toppings in Aisle 11, but on this aisle, we have the type that comes in an aerosol can. There's even a light version. Here is another example of how a shift in mindset is necessary in order to bring about lifestyle changes that stick. Begin weeding these added fats out of your diet, and pretty soon, they simply won't be an issue anymore—they'll be out of your life for good.

A man sidled up to me at the supermarket while I was checking out the flavored, nondairy coffee creamers. He grabbed a carton of amaretto-flavored creamer and, with a grin on his face, asked me if it was legal. He proclaimed it addictive. "You've gotta try it."

I picked up a carton of Irish-cream-flavored creamer, and it said the product was fat free. The ingredient list read: water, sugar, corn syrup solids, sodium diglycerides, artificial flavors, mono- and diglycerides, artificial color, salt, dipotassium phosphate, sodium stearoyl lactylate, xanthan gum, carrageenan, and lecithin. My reaction to that: "Yuck." But there were only thirty calories in a tablespoon, no fat, no cholesterol, no fiber, no protein, no calcium, no iron. . . . A nonfood, I thought to myself.

A product such as this one might be acceptable to some people, but I would treat it the way I would treat the nonfat cream cheese and nonfat sour cream noted earlier. A small amount won't hurt, but I generally recommend avoiding it.

CANNED ROLLS AND BISCUITS, COOKIE DOUGH, AND PIE CRUSTS

Hmmm—what do we have here? This is where you'll find those rolls of ready-made dough in card-

---

Talking shopping with Martha Rose Shulman, author of several excellent cookbooks including *Fast Vegetarian Feasts* and *Mediterranean Light*, Berkeley, California:

Nonfat yogurts and cottage cheeses are getting better all the time (I especially like Colombo yogurt and Knudsen cottage cheese). I don't approve of the many nonfat or low-fat manufactured foods that are full of sugar, additives, and gums (for example, mayonnaises, salad dressings, cream cheeses, and cheeses). They aren't real foods.

board cans—you simply hit the can on the side of the kitchen counter, open it up, and put the dough pieces on a baking sheet. There are cinnamon rolls with five or six grams of fat apiece; some with nine grams apiece. White-flour biscuits with seven grams of fat apiece—and that's without the margarine that many people are in the habit of spreading on. I saw a few acceptable products, such as a French bread dough that only contained one gram of fat per chunk of bread and a quick pizza crust with two and a half grams of fat per piece. Those are fine. A few others were also similarly low in fat, and although they were made with white flour, they would be fine for a change of pace.

As expected, the pie crust was high in fat. The cookie dough contained five to seven grams of fat for two midget-sized cookies—eat at your own risk and only if you think you can stop after two cookies!

Tortillas, English Muffins, and Salad Shells

You may have been wondering when we were going to find more great choices. Well, here's one: whole wheat flour tortillas are a great choice, and they're so versatile! Sure, you can fill one with mashed beans, chopped veggies, and salsa and make a burrito. But you can also spread them with any of your favorite, low-fat sandwich fillings.

To make one of my favorite quick lunches, I spread two whole wheat tortillas with hummus (I make my own low-fat version with pureed garbanzo beans, lemon juice, garlic, and white pepper), roll them up, and maybe eat a piece of fruit with it. This is a good snack or quick and portable lunch.

Do note that some whole wheat tortillas are relatively high in fat, though, so be sure to read the labels. Some are made with shortening and can contain four or five grams of fat apiece. The great choices are the ones that are lower in fat. White-flour tortillas are usually also low in fat, and they're a good choice, too.

Corn tortillas are also good and can be used to make bean and spinach enchiladas and other Mexican-style low-fat dishes. Watch out for the corn tortillas that have been fried, though. Those corn tortilla salad shells can have as much as twelve grams of fat apiece.

You'll find English muffins here, as well. They're typically low in fat and are a reasonable choice. A little further on, we'll find more English muffins mixed in with the other bread products. I sound like a broken record by now, but it's worth repeating: whole grain is best, while regular English muffins can be fine on occasion as well.

Sometimes the ingredients in a product can be great, but the serving suggestions listed on the package can lead you down the wrong path. Serving suggestions listed on one package of sourdough English muffins that I picked up said, "English muffins are marvelous . . . topped with butter or margarine, lavished with honey or tangy marmalade, sprinkled with cinnamon and sugar. Enjoy them served with cream cheese, jelly or jam, garlic or herb butter."

Go ahead and lavish with honey and marmalade, sprinkle with cinnamon and sugar, or serve with jelly or jam. But skip the other fatty suggestions.

HELPFUL HINT

To help ensure a good diet, aim for eating at least six servings of a grain product per day, and choose whole grains most of the time. A serving equals one slice of bread, a bowl of cereal, a half cup of rice, pasta, or hot cereal, one tortilla, and so on.

Eggs

Whole eggs are found here, and refrigerated egg substitute is here as well (we looked at frozen egg substitute on Aisle 11). Whole eggs are loaded with artery-clogging cholesterol. If you must have eggs, then use only the whites. I also suggest that you limit egg whites to two to four per day. The reason again: you don't need the extra protein, and too much animal product in the diet simply displaces fiber-rich, healthful plant matter. Earlier, I mentioned other foods that can substitute for eggs in cooking. If you prefer to use fresh eggs, then use two egg whites to replace one whole egg in recipes that call for eggs. Pitch the yolks.

Breads and Rolls

It's nice to wind up a supermarket tour in an area with so many great choices. Part Two of this book began in the deli/bakery area, where a large variety of interesting, low-fat breads can be found. On Aisle 12, the breads are mostly commercially baked whites, whole grains, and raisin breads, with a few odds and ends in between. Most or all of them are packaged in bread bags that list ingredients and nutrition information. There are lots of good ones from which to choose.

Here you'll have no problem finding a number of whole grain loaves. The first step to picking out a good loaf of bread is to read the list of ingredients and choose a loaf in which a whole grain flour is listed as the first ingredient. Then, scan the remainder of the list, and check the nutrition label to see how much fat the bread contains. Most commercial loaves of bread contain some added oil, but very little. About the only fat-free bread choices to be found in the supermarket are pita bread, some flour tortillas, Italian or French bread, and flat breads or matzo.

A good low-fat bread will still contain one to three grams of fat per slice. That's okay, but compare loaves. Some are made with more whole grain flour than others, and some contain more fiber than others. It may be better to choose a bread with two grams of fat per slice that is made with 100 percent whole grain flour than to choose one with only one gram of fat and more refined flour. Take a look at the following examples:

stone-ground 100 percent whole wheat bread

Ingredients: stone-ground 100 percent whole wheat flour, water, high-fructose corn syrup, wheat gluten, yeast, honey, salt, molasses, partially hydrogenated soybean oil, raisin syrup, ethoxylated mono- and diglycerides, vinegar, soy flour

Serving size: 2 slices
Calories: 100
Protein (gm): 5
Total fat (gm): 1.5
Cholesterol (mg): 0
Carbohydrate (gm): 20
Dietary fiber (gm): 3
Sodium (mg): 170
Calcium (% DV): 0
Iron (% DV): 4

100 percent whole wheat bread

Ingredients: stone-ground whole wheat flour, brown sugar, yeast, wheat gluten, contains 2 percent or less of each of the following: honey, salt, vegetable oil (soybean oil or canola oil), dough conditioners (sodium stearoyl-2-lactylate, monoglycerides, ethoxylated mono- and diglycerides, calcium iodate, calcium peroxide), cultured whey, vinegar, calcium sulfate, monocalcium phosphate, yeast food (ammonium sulfate)

Serving size: 1 slice
Calories: 60
Protein (gm): 4
Total fat (gm): 1
Carbohydrate (gm): 11
Dietary fiber (gm): 3
Cholesterol (mg): 0
Sodium (mg): 125
Calcium (% DV): 4
Iron (% DV): 2

natural wheat bread

Ingredients: ground whole wheat, water, unbleached enriched wheat flour (flour, malted barley, niacin, reduced iron, thiamine mononitrate [vitamin B₁] and riboflavin [vitamin B₂]), high-fructose corn syrup, partially hydrogenated soybean oil, salt, yeast, vinegar	Serving size: 1 slice Calories: 80 Protein (gm): 3 Total fat (gm): 1 Carbohydrate (gm): 17 Dietary fiber (gm): 2 Cholesterol (mg): 0 Sodium (mg): 200 Calcium (% DV): 0 Iron (% DV): 4

The first thing you might notice when comparing the labels on these three loaves of bread is that the serving size for the first one is listed as two slices, whereas the serving size for the others is listed as one slice. So in order to compare nutrition information, divide the figures given for the first bread in half.

With that done, you will see that the fat and fiber contents of the three loaves are fairly similar. The first loaf has slightly less fat and about half the fiber of the other two loaves. The second loaf has the least fat and the most fiber of the three. All are cholesterol free; the first loaf has the least amount of sodium (remember, the information given has to be divided in half).

Looking at the ingredient listings, all three list a whole grain flour as the first ingredient. All contain a small amount of vegetable oil. The first and third loaves contain fewer additives, such as dough conditioners.

The best choice? All things considered, any of these three would be a good choice. My personal choice would be the third loaf, because it is low in fat, made with all whole grain flour, and it has a nice, short ingredient list without unnecessary additives. Considering that these three examples are fairly similar in nutritional content, you might also

> **HELPFUL HINT**
>
> When choosing bread, pay more attention to the ingredients—is it made with whole grains?—and the fat content, and don't worry so much about calories. "Light" and "thin-sliced" bread is marketed for its lower calorie content (playing on the myth that bread is fattening), a far less important detail than those just mentioned. The only reason to buy thinly sliced bread is if you just happen to prefer it that way.

choose one over another based upon taste preference or differences in texture. The first and second loaves are softer, fluffier breads; the third is a denser, heavier loaf.

Hamburger, hot dog, and other rolls typically are made with white flour, but if you look, you'll probably also see at least one brand that makes a whole grain version. Finally, in the midst of all the breads, you'll once again see doughnuts, snack cakes, and more desserts. A quick peek at a loaf of pound cake showed that one slice—one *small* slice—had ten grams of fat. A better choice is angel food cake, which is virtually fat free and can be topped with fruit. We saw angel food cake first in the deli/bakery area.

Cones and Toppings

Cones and toppings are familiar accessories to ice cream (coming up next). By themselves, they can be fine. The old standard, light-as-a-whisper yellow waffle cone is pretty benign by itself. And there are many toppings now that are fat free. I saw some fat-free hot fudge sauce that was made with fairly wholesome ingredients, fruit toppings such as strawberry and pineapple, and some caramel and butterscotch toppings that contained only one gram of fat in two teaspoons of topping (the artificial color and flavor that they contained put me off, though). All of these are fine as once-in-a-while treats—like any low-fat sweets. What matters more is what you fill the cone with and what you pour the topping over.

Ice Cream, Frozen Juice Bars, Italian Ices, and More

Sure, ice cream is dripping with fat and cholesterol, and lately, with some of the new flavors that some of the premium brands are putting out, it seems as though it's a battle to see who can pile on the most butterfat. But there are also lots of great choices here.

We've talked about sweets throughout this part of the book, and I've generally given the thumbs-up to nonfat sweets that are made with wholesome ingredients (and a minimum of undesirable additives, artificial colors, and artificial flavors)—in other words, a *qualified* thumbs-up. Since sweets are typically empty-calorie foods, they should be limited to not more than two per day. With that said, there are plenty of choices here that are perfectly acceptable and can take the place of their unhealthier cousins. Take a look at a couple:

Talking shopping with Robert Pritikin, director of the Pritikin Longevity Centers, Santa Monica, California:

I like to take overripe bananas and freeze them. Then take them out of the freezer and eat them. They're creamy and really good. That was the only ice cream I got as a kid.

Italian ice

Ingredients: water, sugar, corn syrup, concentrated strawberry juice, citric acid, natural flavor, guar and cellulose gums, red beet color	Serving size: 1 6-ounce cup
	Calories: 100
	Protein (gm): > 1
	Total fat (gm): 0
	Cholesterol (mg): 0
	Carbohydrate (gm): 26
	Dietary fiber (gm): 0
	Sodium (mg): 0
	Calcium (% DV): 0
	Iron (% DV): 0

paletas (watermelon)

Ingredients: watermelon, water, sugar, natural color	Serving size: 1 2.5-ounce bar
	Calories: 57
	No other information available

As you can see from the nutritional information, Italian ices are not exactly nutritional heavy hitters, but they're made with fairly wholesome ingredients—even the food coloring in this example is natural beet color. As a treat, these ices are a great choice.

My personal favorites are the paletas, which look like large-sized Popsicles and come in a wonderful variety of absolutely delicious fruit flavors—lime, pineapple, banana, strawberry, watermelon, and so on. The watermelon variety comes complete with the seeds, which look pretty. These paletas contain just fruit and sugar. They're a great choice.

I prefer the two examples given here to most of the commercial Popsicles that are available, mostly because of all of the artificial flavorings and colorings that other products contain. Some are sweetened with aspartame, too, which, again, I personally prefer to avoid. There are also some juice bars available that are made with natural flavorings and colorings and are sweetened with sugar rather than aspartame, and I think that these are a better choice than many Popsicles.

Back to ice cream. There is also a sea of other ice cream treats—dessert bars, ice cream sandwiches, ice cream candy bars, and more. Some are sweetened with aspartame, some with saccharin. Some fat-per-servings totals are in the double digits. And if you spy words such as "reduced fat" on labels, take the time to check out the facts. One reduced-fat candy-bar ice cream bar had seven grams of fat in one teeny-tiny two-and-a-half ounce bar (compared to fourteen grams of fat in the regular, three-ounce bar version). If you're like me, a two-and-a-half-ounce bar would be three bites, and I'd want about six more (bars, that is).

There are some yogurt bars with less than one gram of fat per serving, although I noticed that these tended to contain artificial colors. There is some frozen nonfat yogurt, which can be an acceptable choice as well. You may also see a nondairy ice cream product that contains no cholesterol or saturated fat; check the nutrition information. The one I looked at contained eleven grams of fat in a half-cup serving. Cholesterol or no cholesterol, that's too much fat. Sherbets vary in fat content; some contain cream and butterfat, whereas some contain only a small amount of milk. With all

---

HELPFUL HINTS

When you choose a frozen treat, consider the following points:

- Go for the all-fruit options first, especially those made with whole fruits (they'll contain some fiber).
- Choose products with natural ingredients (colorings, flavorings, sweeteners, and so on).
- Next best are nonfat dairy options, again with natural ingredients.

of the other nonfat frozen desserts available, it is possible to find something that tastes good and is actually fat free.

The Last Bite: Ready-made Mashed Potatoes and Orange Juice in Cartons

You'll see fresh, ready-made mashed potatoes in the refrigerated section of Aisle 12. I recommend that you pass on these. The ones I've seen contain undesirable preservatives, for starters. Take a look at one of these products:

ready-made mashed potatoes

Ingredients: potatoes, margarine, salt, disodium dihydrogen pyrophosphate (to retain color), potassium sorbate, sodium bisulfite	Serving size: ⅔ C Calories: 110 Protein (gm): 3 Total fat (gm): 2 Carbohydrate (gm): 19 Cholesterol (mg): 0 Dietary fiber (gm): 1 Sodium (mg): 210 Calcium (% DV): 2 Iron (% DV): 4

A similar hash brown product contains only a half gram of fat per half-cup serving, but like the mashed potato product, it contains preservatives such as sulfites.

This product would be exceptionally easy to make at home from fresh potatoes, and you could leave out the added fat and salt, as well as doing without the preservatives. Even though this product is reasonably low in total fat and contains no cholesterol, I'd still skip it, all things considered. Every once in a while? Okay.

This is also the aisle where you'll find the fruit juices in refrigerated cartons. Is one variety better than another?

That depends. The freshly squeezed juices that are usually sold in the produce section taste so superior to any other juices that it's hard to believe that they aren't nutritionally superior as well. But their cost is prohibitive for many people.

So some folks go for the carton juices instead, some of which are not from concentrate. I'm not sure that anyone knows the degree to which processing (and pasteurization) affects the nutritional value of carton juices. Some are now fortified with vitamins A, C, E, thiamin, and niacin; others are fortified with calcium. The calcium-fortified orange juices can be helpful for people who don't want to use dairy products but who want to boost their cal-

---

GREAT CHOICES

- fresh pasta made with egg whites (no yolks) or egg free
- fat-free or very low-fat fresh tomato sauce
- whole grain, low-fat pizza crusts
- ready-made rolls of bread dough with two or fewer grams of fat per piece (buy whole grain when you can)
- flour tortillas, especially whole wheat
- corn tortillas (not fried)
- English muffins
- whole grain breads and rolls, low fat
- angel food cake with fruit as a treat
- Italian ices
- paletas
- Popsicles—naturally flavored and colored
- frozen fruit-juice bars—naturally flavored and colored
- nonfat frozen yogurt and nonfat yogurt bars

---

cium intake the easy way—by gulping down a beverage instead of cooking broccoli.

Aside from that, I would simply apply the same criteria to choosing juices as were discussed in Aisle 11. Remember, don't completely replace fresh fruit in your diet with fruit juice—you need the fresh fruit for its fiber and for the vitamins and minerals it contains in its unprocessed form.

PART
THREE

Putting
It
All
Together

What's Your Shopping Style?

I hear it every day: after "What's left to eat?" another frequent question that I'm asked about putting dietary recommendations into practice is "Where do I begin?" Well, just having a grocery list in hand can be a big help.

But not everyone likes to shop from a list. Some people are more comfortable just perusing the aisles and picking up items that they know they need or that strike them as a good idea when they see them. I, on the other hand, invariably forget an item or two that I need if I don't keep a list. Who wants to make an extra trip to the supermarket because you forgot to pick up orange juice?

People have different shopping styles that suit their individual lifestyles and personalities. Some people, for instance, like to plan their meals a week at a time. Then they draw up shopping lists that correspond to their menus. Planning ahead in this way can also be helpful for people who are learning new eating skills, since it gives them more control over their meals. Planning ahead helps to prevent impulsive decisions about meals and increases the probability that you will make good food choices.

Other people like to have a variety of staples in

the house and simply draw from whatever is on hand when they fix a meal. They may decide what they are going to make a day, an hour, or even minutes before a meal, using whatever is available and depending on their mood. Assuming what's on hand is healthful, then this approach can be fine, too. Do what works for you.

Most of the time, my own personal style is to decide what to make less than an hour—and sometimes only minutes—before a meal. I fix what's on hand and whatever I'm in the mood to eat. My meals are simple and often take only minutes to prepare. The key for me is to keep lots of fresh fruits, veggies, whole grain rolls, breads, bagels, and cereals on hand. I keep a piece of scrap paper on the refrigerator, and whenever I notice that I am low on an item, I add it to my running grocery list. Then I take that list with me when I shop.

You can, of course, compose your own grocery lists from the lists of great choices given throughout Part Two. Some people will prefer the ready-made lists that follow. You can revise these somewhat to suit your own food preferences or buying habits. If you prefer a different format, rewrite them onto separate pieces of paper.

If you like them just the way they are, you may want to make photocopies of the lists in this book. Tape them to the refrigerator door to help you keep a running inventory and then carry them to the supermarket when you go shopping. These lists are versatile. There are separate lists for foods that you are likely to purchase weekly, monthly, and less often.

Sample Grocery Lists

The following lists are adapted from ones that I wrote for *Simple, Lowfat & Vegetarian* (Baltimore,

Md.: Vegetarian Resource Group, 1994) and are reprinted here with permission from the publisher.

WEEKLY SHOPPING LIST

These are foods that you are likely to purchase more frequently, because they are perishable and will only keep in your refrigerator for a few days to a week.

fresh fruit (especially locally grown, in season)
> apples
> apricots
> bananas
> blueberries
> cantaloupe
> cranberries
> grapefruit
> grapes
> guavas
> honeydew
> kiwi
> lemons
> limes
> mangoes
> nectarines
> oranges
> peaches
> pears
> pineapples
> plums
> strawberries
> watermelon
> others

prepared fresh fruits
> fresh juices: orange, grapefruit, tangerine, apple cider, others
> chilled, bottled mango slices, pineapple chunks, fruit salad
> packaged, cut fruits

fresh vegetables (especially locally grown, in season)

- asparagus
- bean sprouts
- beets
- bell peppers
- bok choy
- broccoli
- brussels sprouts
- cabbage
- carrots
- cauliflower
- celery
- collard greens
- corn
- cucumbers
- kale
- leeks
- mustard greens
- onions
- potatoes
- tomatoes
- others

prepared fresh vegetables

- fresh vegetable juices: carrot, carrot spinach, beet
- packaged, cut vegetables and mixed greens
- fresh herbs: basil, dill, mint, rosemary, sage, thyme and others

breads (especially whole grain)

- bread sticks
- dark rye
- English muffins
- French bread
- German whole grain
- Italian bread
- multigrain
- oatmeal

    pita pockets
    pumpernickel
    rye
    sourdough
    whole wheat
    others

rolls (especially whole grain)
    garlic knot
    hard rolls
    hoagie rolls
    kaiser rolls
    onion
    pumpernickel
    rye
    whole wheat
    others

bagels (avoid egg)
    apple raisin
    blueberry
    cinnamon raisin
    mixed grain
    oat bran
    onion
    plain
    poppy seed
    pumpernickel
    rye
    sesame seed
    whole wheat
    others

other bread products
    corn tortillas (not fried)
    flour tortillas (especially whole wheat)
    fresh whole wheat pizza crusts
    low-fat whole grain muffins

fresh pasta (made without egg yolks)

fresh deli items
    four-bean salad (oil free)

fresh fat-free salad dressings
fresh marinara sauce
fresh pizza (no cheese; marinara sauce with
  veggie toppings)
fresh salsa
sauerkraut
others (fat free)
angel food cake (for dessert or a treat)
skim milk, nonfat yogurt, nonfat cheese (if
  desired; limit)
egg whites (if desired; limit)
fresh tofu (if desired; buy low fat)

Also need:

_____

_____

_____

_____

_____

_____

_____

_____

MONTHLY SHOPPING LIST

These items can be purchased less often, since
they keep longer in the cupboard or freezer.

CUPBOARD STAPLES

canned goods

beans (canned or dry): baked beans (vegetarian, preferably fat free), garbanzo, black,
  pinto, kidney, navy, split peas, lentils, and
  others
bean salad
artichoke hearts (no oil added)
soups (fat free): lentil, vegetable, vegetarian
  split pea, tomato
pasta sauces (all fat free)

tomato sauce and paste

applesauce

fruits: peaches, pears, mandarin oranges, grape-fruit sections, cranberry sauce, fruit cocktail, apricots, pineapple

vegetables (no fat added): green beans, peas, carrots, asparagus, corn, tomatoes, and others

sloppy joe sauce (fat free)

refried beans (preferably fat free)

dry items

pasta (made without egg yolks)

dry cereals (whole grain, low fat): raisin bran, shredded wheat, bran flakes, and others

hot cereals (whole grain): oatmeal, whole wheat, others

whole wheat flour

whole grain bread, pancake, and all-purpose mixes (modify to omit added fat)

other whole grain mixes (modify to omit added fat)

rice: basmati, brown, wild, and others

barley

bulgur wheat

couscous

instant mashed potatoes (no added fat)

soup mixes (no added fat)

vegetable oil spray

snacks and treats

rice cakes and popcorn cakes

popcorn (fat free or very low fat)

other fat-free snacks (chips, pretzels, granola bars, whole grain and fat-free toaster pastries)

bean dip (fat free)

crackers (fat free, preferably whole grain)

flat breads (including matzo) and bread sticks

cookies (treats; fat free, preferably whole grain)

candies (treats; fruit-flavored sours, hard candies, and pectin jelly beans)

tapioca and flan (treats; make with nonfat
milk or low-fat soy milk)
canned pumpkin and fruit pie fillings
fruit toppings: prune, cherry, apricot, and others

flour tortillas (especially whole wheat) and corn
tortillas

low-fat tofu and soy milk (in aseptic packages)

condiments
fat-free salad dressings
fat-free mayonnaise
horseradish mustard
salsa
ketchup
fruit preserves and conserves
fruit-only or low-sugar spreads
syrups, molasses
sun-dried tomatoes
grape leaves in brine
hoisin sauce
stir-fry sauces (fat free)
low-fat marinades and barbecue sauce
mustard
pickles and pickle relish
vinegar: balsamic, herbed, fruited, and others
jams and jellies
honey
chutney
sweet-and-sour sauce
spicy brown bean sauce
natural, butter-flavored sprinkles

herbs and spices
basil
bay leaves
cinnamon
cumin
curry powder
dill
fat-free vegetable bouillon

    garlic
    ginger
    paprika
    pepper
    others

natural sodas

water
    plain or flavored mineral waters
    plain or flavored seltzer waters
    club soda

bottled fruit juices and blends including sparkling cider and sparkling grape juice

bottled vegetable juices
    tomato
    V-8
    borscht
    carrot

herbal tea

dried fruits
    apples
    apricots
    cherries
    currants
    dates
    figs
    mixed fruits
    prunes
    raisins

FREEZER STAPLES

frozen bagels (avoid egg)
    blueberry
    cinnamon raisin
    oat bran
    onion
    plain
    others

muffins and dinner rolls (whole grain, low fat)

frozen waffles and pancakes (whole grain, low fat)
frozen pasta (made without egg yolks)
frozen juices (avoid coconut milk)
  apple
  cranberry raspberry
  grape
  grapefruit
  lemonade
  limeade
  orange
  pineapple orange
  others
frozen entrées (fat and cholesterol free or very low fat)
meat substitutes (fat free or very low fat)
  veggie burger patties
  veggie hot dogs
  others
frozen novelties
  frozen juice bars
  Italian ices
  nonfat frozen yogurt and nonfat yogurt bars
  paletas
  Popsicles
frozen egg substitute (if desired; limit)
frozen fruit
  berry mixtures
  blueberries
  cherries
  mixed fruit
  peaches
  pineapple
  red raspberries
  strawberries
plain frozen vegetables
  broccoli
  spinach
  cauliflower

cut corn
green beans
mixed vegetables
lima beans
green peas
carrots
stir-fry mixes
others

Also need:

_____

_____

_____

_____

_____

_____

_____

_____

Grocery Cart Makeovers

How many times have you glanced into the cart in front of you at the checkout line and evaluated what you saw? Maybe it's just dietitians who do that. I do it. I give other people's grocery carts mental makeovers. When I see a loaf of soft commercial white bread, I think, *I'd trade that for some heavy, coarse-grained whole wheat bread*. Or I attribute great qualities to the person pushing the cart containing piles of fresh fruits and veggies (*Gee, that person must be pretty smart; I wonder where he learned so much about nutrition.*).

Maybe this is my own brand of zeal, but it's a great exercise. The carts that follow are real-life examples of people's choices. On one Saturday night (no, I had nothing better to do), I staked out the checkout lines at Harris Teeter's flagship Morrocroft store in Charlotte, North Carolina. I positioned myself between two "ten items or less" lanes and a regular lane. Yes, I looked very weird, standing behind the cashiers, surreptitiously recording people's purchases with my hand-held tape recorder. I drew some suspicious stares. One man moved to another line.*

*Nutrition students: this makes a great class project, but if you plan to give it a shot, be sure to let the store manager and

How would you make over the carts that follow? What would you do differently?*

To get you started, I've used the first two baskets as examples. Evaluate the others on your own first, then take a peek at my suggestions, if you wish.

Cart One (woman, early seventies)

two boxes of unsalted matzos
two boxes of melba toast, one regular and one whole wheat
one large jar of Smucker's strawberry preserves
one box of Uncle Ben's white rice
half gallon of Crowley's frozen yogurt
half gallon of Colombo frozen yogurt

What I'd Do Differently

I'll bet this woman eats the preserves on the matzos and melba toast as a snack. Not bad, if you have to have a vice.

The frozen yogurt should be nonfat, and it should be just a once-in-a-while treat. Since she is purchasing two half gallons, she's probably snacking a bit liberally on it. She might consider frozen fruit bars or nonfat fruit sorbet as an alternative to nonfat frozen yogurt.

The matzos and melba toast are fine. Whole

---

cashiers know what you are doing. It doesn't hurt to be sociable and talk to the customers as they come through the line, too, casually letting them know what you are up to. Take my word for it; it'll reduce your embarrassment.

* Granted, this was a Saturday night and "party time" for some folks. Depending on "where you are at" in your own dietary evolution, you might look at some of these baskets and say, "Well, I guess I'll let that item get by this time—it *is* Saturday night, after all." Others will opt for healthier choices. Do what's right for you. Not only does this exercise give you practice making food choices; it also may prompt some thoughts about your personal dietary goals—about where you are now and where you still want to go.

wheat matzos are also available and would be a great choice. Of course, brown rice is also available, but white rice is also acceptable. Remember, choose whole grains as often as possible.

Cart Two (two teenage girls)
three fresh, ready-made pizzas—one cheese, two with pepperoni

What I'd Do Differently
These were picked up in the deli area, and if they were made to order, I would have asked for cheeseless, veggie pizzas. Another option (especially if quick-and-easy was the goal): pick up ready-made whole wheat crusts, some fresh tomato pasta sauce from the deli (or canned sauce), and some presliced veggies for the topping.

Cart Three (man, midthirties)
two pork chops
one can of green peas
one package of fresh shrimp
one package of Pronto potatoes (home fries)

Try your hand at this one and those that follow. Afterward, take a peek at my suggestions, located at the end of this section, if you wish.

A man one aisle over motions for me to come over. "Are you with the Secret Service?"

"No, I'm a dietitian working on a personal supermarket project."

"Oh. Well, what do you think of what I'm buying here?"

Cart Four (woman, midforties)
one frozen pizza with sausage and pepperoni
three 2-liter bottles of cola (one diet)

Cart Five (woman, midfifties)

bananas
grapes
loose bagels
Breyer's regular yogurt
one pound of Fleishmann's margarine
half gallon of Tropicana orange juice
quart of 2% low-fat milk
small container of regular cottage cheese

Cart Six (couple, midthirties)

loaf of French bread
one eggplant
mushrooms
scallions
one sweet potato
container of sour cream
bag of self-rising cornmeal mix
bag of regular tortilla chips
one box of specialty ice cream loaf

Cart Seven (woman, midsixties)

one twelve-pack of Coke
four boxes of Quaker Toasted Oatmeal honey nut
  cereal
one bag of Fresh Express mixed salad greens
half gallon of buttermilk
one pound of bacon
one frozen Lean Cuisine spaghetti dinner with
  meat sauce

At this point, a young female cashier wanders over and, assuming that I am a nutrition student, says that she really admires how I am getting real-life examples of what people buy. "If I had to do it, I'd probably just have stayed home and made it up."

Cart Eight (two women, late twenties)
twelve-pack of light beer
twelve-pack of "ice" beer
bottle of cola
bottle of ginger ale
bottle of Quibell mineral water
large bag of pretzels
large bag of potato chips
large bag of regular tortilla chips
jar of Taco Bell cheese sauce
container of sour cream
pack of cigarettes (we're in North Carolina)
Hershey's chocolate bar

Cart Nine (couple, midfifties)
chunk of Stilton cheese
box of specialty crackers (round, white-flour
   type)
bottle of port

Cart Ten (woman, late forties)
box of German chocolate cake mix
one-pound can of honey-roasted peanuts
round loaf of white bread
pound of Promise margarine
carton of light sour cream
half dozen eggs
one onion
bottle of reduced-fat mayonnaise
12-ounce bag of chocolate chips
bag of hard candies
small bottle of vegetable oil
bottle of dried basil
bottle of white wine

One man asks why I can't just keep people's cash register receipts, since these list what was purchased (a scanner
is used at the checkouts). I explain that the store needs to

give people their receipts. Also, I need more detail about some of the products than is provided on the receipts.

This man has given me an idea, though. It might be worthwhile for some folks to hang on to their cash register receipts and to review them from time to time as a way of checking buying habits. Just as some people like to keep food diaries when they are watching their eating habits, cash register receipts can be good diaries of shopping habits.

The receipts may not give all of the detail about certain products, but the chances are good that with the receipt as a clue, you will remember the details about what you bought. (Was it nonfat cheese or regular, for instance?)

Cart Eleven (woman, late thirties)
large bag of regular tortilla chips
two bags of frozen Sara Lee plain bagels
package of imitation crabmeat
small frozen cheese pizza
package of Entenmann's fat-free chocolate
  brownie cookies
container of Philadelphia Light Cream Cheese
jalapeño and cheddar cheese dip
container of nonfat yogurt
one pound of regular cheddar cheese
six-pack of Coke
six-pack of beer
two gallons of drinking water

Makeover Suggestions for Carts Three Through Eleven

Cart Three
This looks like someone's interpretation of surf and turf. My guess is that these foods were

being purchased as a meal for one or two people. It's a typical American meat-centered meal, where meat is the focal point of the plate and the vegetables are relegated to a minor role. The meal is low in fiber and too high in fat and protein.

A makeover should include role reversal at the very least, using the meat as a condiment or minor ingredient and loading up on vegetables and grains instead. Skip the pork chops, for instance, and if you must have some meat, then use a few pieces of shrimp, either plain on the side or incorporated into a salad or vegetable dish.

You could also omit the meat altogether. You could add some additional vegetables to the peas—carrots, onions, yellow squash—to make an interesting mixture. Heat the potatoes in a nonstick skillet without added fat, or bake them in the oven. A hearty bowl of lentil soup, a small green salad, and/or a whole grain roll would make this a satisfying meal.

Cart Four

I would treat this example in much the same way as Cart Two. Avoid the sausage and pepperoni. Either choose a plain frozen pizza (looking for the choice lowest in fat) and add your own vegetable toppings, or buy a ready-made whole wheat crust and make your own vegetarian pizza. Leave the cheese off or use a small amount of grated nonfat cheese.

Team this with a quick fruit salad made by slicing and tossing together whatever fresh fruit you have on hand—a banana, a pear, an orange, a kiwi. I'd opt for a tall glass of sparkling mineral water with a twist of lemon or lime in place of the cola.

Cart Five

The bananas, grapes, and bagels are great choices. The orange juice is fine, too. This cart could be improved by choosing nonfat yogurt and cottage cheese in place of the regular varieties, and skim milk should be substituted for 2%. As for the margarine—add as little fat to foods as possible. Make this pound of margarine last six months.

Cart Six

A few simple changes would improve the nutritional profile of this cart of groceries. Instead of regular tortilla chips, I'd choose nonfat baked chips (and eat them with salsa). In place of the ice cream, I'd substitute nonfat frozen yogurt or fruit sorbet. Assuming the sour cream was being used as a condiment, it could be replaced with nonfat plain yogurt or fat-free sour cream.

Cart Seven

This cart, on the other hand, needs an overhaul. Replace the cola with mineral water (as in Cart Four) or seltzer water, or mix a little fruit juice with some sparkling water for a change of pace. Refined grain cereals won't hurt once in a while, but choose whole grain cereals more often, such as raisin bran, bran flakes, shredded wheat, and others.

Buttermilk is usually made from 2% or whole milk. If it's used in small quantities for baking, low-fat buttermilk is fine. However, if it's being used in larger quantities as a beverage, then skim milk should be chosen instead.

Avoid bacon. If you must have a slice or two occasionally, then I suggest buying one of the meatless, vegetable protein varieties that can easily be found in supermarkets and natural foods stores. If you want a frozen entrée, find one with the least amount of fat and cholesterol. Add a green salad and some whole grain bread to the meal; have fresh fruit for dessert.

Cart Eight

Whoa! I would salvage the pretzels and mineral water from this cart. It's Saturday night, so this must be party fare. If I were heading to this party myself, I'd bring along a supply of fresh cut vegetables, some whole grain crackers, salsa, and nonfat black bean dip. I'd bring plenty, too, because I'd have to share. There's a high demand for healthy foods at parties.

Cart Nine

Fatty cheese on refined-flour crackers with a glass of port. Maybe this is a tradition that this couple grew to love. My suggestion is to cultivate new traditions that are health supporting and that can eventually take the place of the old, less desirable ones. A piece of exotic fruit, for instance, with a glass of wine could be just as romantic or enjoyable as a chunk of cheese on a cracker.

Cart Ten

More Saturday night entertaining, no doubt. Most of these groceries appear to be supplies for baked desserts. A few of the ingredients chosen are lower-fat versions of fatty foods, such as sour cream and mayonnaise. Some reduced-fat products, such as light margarine, do not work well in baked goods and cannot be used.

Any recipe could be further modified by using only the whites of the eggs and throwing away the yolks (two egg whites equal one whole egg in recipes). The amount of oil used could also be reduced by one-third to one-half in most recipes. In most baked goods, a scoop of applesauce or canned pumpkin can also be substituted for all or some of the fat. If you do make baked desserts, I suggest reducing the fat wherever feasible and even using part whole grain flour when possible.

In this example, the ingredients in the cart are probably destined to be used in low-fiber, high-fat foods and dishes. Once in a while these can be okay, but since once in a while happens too frequently for many of us, I suggest looking for ways to replace some of these foods with healthier options.

Cart Eleven

We've covered such items as regular tortilla chips, bagels, soft drinks, full-fat dairy products, and frozen pizza in the previous examples. The imitation crabmeat could be treated like the shrimp in Cart Three—use it as a condiment or skip it altogether. I would replace the jalapeño and cheddar cheese dip with a nonfat bean dip and salsa. The fat-free cookies are fiberless, but they could be fine if eaten only a couple at a time, as a treat. Likewise, the beer could be acceptable if it is consumed in small quantities—not more than one drink a day, two at the most.

This cart, like some of the others before it, looks like it's holding weekend snack foods. I would boost the nutritional contribution by throwing in some fresh fruits and vegetables that could be cut up and served as finger foods or tossed in salads. Having foods like these on hand and ready to eat is one of the secrets to eating well.

A Few More Words About Supermarket Shopping . . .

Robert Pritikin, director of the Pritikin Longevity Centers, Santa Monica, California:

Before we talk about what to put into the basket at the store, we have to think about cleaning your house out first. The key is not to have anything in the house that is bad for you. It's also critical to have family support; you have to make a decision as a family unit to eat a low-fat diet. Don't fool yourself into thinking that you'll buy different things for your kids than you do for yourself. If you buy ice cream and treats for your kids, you'll eat it.

You are weak when you go through a transition; you're fragile. Lifestyle changes cause discomfort. You need support or the frustrations that come with making a lifestyle change will lead to a relapse. Don't try to make all the changes overnight. My recommendation is to take a step at a time.

First, analyze your situation. Look for the changes you can make. That's number one.

Number two: plan the changes that you'll make step by step starting from where you're currently at. Don't make so much discomfort for yourself that you collapse. First deal with making sustainable changes, then you'll get sustainable results.

Number three: create a supportive environment at home.

Laurel Robertson, coauthor of *The New Laurel's Kitchen*, Tomales, California:

I have to confess that I almost never go into supermarkets—we have a big garden and belong to a bulk food–buying cooperative. Several families and friends collaborate to share cooking and cleanup.

It's really wonderful that supermarkets offer such good things now, and I am happy to know that people like you are helping folks get the best out of them. But I hope that you can put in one word about the bigger picture.

Everyone's busy, and it's great to find nutritious low-fat entrées in the freezer case; do we connect our purchase with the demise of local agriculture and the advancing urban sprawl that oppresses us so much? We are vegetarians out of compassion, but does our appetite for zucchini in December dispossess farmers in some sunny southern land? Soon we will welcome "natural" vanilla for a tiny price—the product of biotechnology tissue culture. Will we hear about the seventy thousand people in Madagascar, the thirteenth poorest country in the world, who *used* to be able to make their living growing vanilla orchids? There is a pattern here that affects all of us, but first and especially, the poorest. I think we need to look for ways to counteract it. Part of the remedy is gardening, seed saving, buying locally, eating seasonally. Supermarket shoppers can play a big part in that by choosing thoughtfully, and by asking store managers to stock and brag about produce that is locally grown and products that are locally produced.

Daily Food Guide

This food guide is an adaptation of two other guides—one that appears in *Eating Well—The Vegetarian Way*, a brochure that I wrote for the American Dietetic Association (1992), and in *The New Four Food Groups*, a vegetarian food guide created by the Physicians Committee for Responsible Medicine (1992).

Food Group	Daily Servings	Serving Sizes
Whole Grains	6 or more	1 slice of bread, 1/2 cup of hot cereal, 1 oz. dry cereal
Vegetables	4 or more	1/2 cup cooked or 1 cup raw
Legumes	2 to 3	1/2 cup cooked beans, 8 oz. low-fat soy milk, 4 oz. light tofu
Fruits	3 or more	1 piece of fresh fruit, 1/2 cup fruit juice, 1/2 cup canned or cooked fruit, 2 T dried fruit
Optional: Nonfat Dairy Products	limit 2	8 oz. nonfat yogurt, 8 oz. skim milk, 1 oz. nonfat cheese

Food Group	Daily Servings	Serving Sizes
Egg Whites	limit 2	2 egg whites
Fats	Added fats should generally be avoided	
Sweets	Try to limit nonfat sweets to one or two servings per day	

Those who consume no animal products should take a vitamin B_{12} (cyanocobalamin) supplement or eat vitamin B_{12}–fortified foods. In addition, if exposure to sunlight is limited, a vitamin D supplement of no more than 100 percent of the RDA may be indicated.

Counting Those Fat Grams

Somewhere along the line, weight watchers replaced calorie counting with fat gram counting. This makes some sense, since we now know that in weight control, it's not so much the number of calories that matters as it is the source of those calories.* Fat is a concentrated source of calories, and some people's bodies are very efficient at storing dietary fat as body fat.

So, counting grams of fat can be a convenient way to keep the diet in check. As the saying goes, "worry about the dollars and not the pennies." For many people, controlling their fat intake can provide the biggest return for their efforts: keep the fat intake low enough, and everything else is likely to fall into place. No need to count anything else.

For those who like structure, keeping a tally of how many grams of fat they are eating each day can

---

*Of course, at some point, calories *do* count. It is possible for some people to eat so many calories—even from fat-free foods—that they cannot lose weight. Generally, however, sticking to a low-fat diet helps control weight, because low-fat foods such as fruits, whole grains, and vegetables are bulky. You are likely to feel full and stop eating before you have consumed an excessive number of calories.

be useful. Just keep in mind, though, that focusing only on fat may distract from the bigger picture. As I mentioned at the beginning of the book, simply replacing high-fat animal products with low-fat or nonfat varieties isn't always good enough. There needs to be an overall shift away from animal products to a diet that is heavily reliant on foods of plant origin in order to produce the most benefit.

If the goal is to hold the fat level of the diet to about 10 to 15 percent of the calories, then most women will probably want to hold their fat intake to not more than about twenty to thirty grams of fat per day. That corresponds to a diet of about 1,800 calories. Most men will do well to hold their fat intake to about twenty-six to thirty-eight grams of fat per day, which corresponds to a diet of about 2,300 calories.

These numbers are just a rule of thumb, though. Some simple calculations can help you individualize your fat gram goals.

Calculating Your Fat Gram Goal

The first step is to decide how much you would like to weigh. There's more than one way to determine that. If you know from experience the weight at which you feel your best, you can simply use that figure. Otherwise, refer to these 1983 Metropolitan Life Insurance charts to get an estimate of your best weight.

1983 Metropolitan Height and Weight Tables
Men

Height Feet	Inches	Small Frame	Medium Frame	Large Frame
5	2	128–134	131–141	138–150
5	3	130–136	133-143	140-153
5	4	132–138	135–145	142–156
5	5	134–140	137–148	144–160

5	6	136–142	139–151	146–164
5	7	138–145	142–154	149–168
5	8	140–148	145–157	152–172
5	9	142–151	148–160	155–176
5	10	144–154	151–163	158–180
5	11	146–157	154–166	161–184
6	0	149–160	157–170	164–188
6	1	152–164	160–174	168–192
6	2	155–168	164–178	172–197
6	3	158–172	167–182	176–202
6	4	162–176	171–187	181–207

Weights at ages 25–59 based on lowest mortality. Weight in pounds according to frame (in indoor clothing weighing 5 lbs., shoes with 1-inch heels.

Women

HEIGHT		Small	Medium	Large
Feet	Inches	Frame	Frame	Frame
4	10	102–111	109–121	118–131
4	11	103–113	111–123	120–134
5	0	104–115	113–126	122–137
5	1	106–118	115–129	125–140
5	2	108–121	118–132	128–143
5	3	111–124	121–135	131–147
5	4	114–127	124–138	134–151
5	5	117–130	127–141	137–155
5	6	120–133	130–144	140–159
5	7	123–136	133–147	143–163
5	8	126–139	136–150	146–167
5	9	129–142	139–153	149–170
5	10	132–145	142–156	152–173
5	11	135–148	145–159	155–176
6	0	138–151	148–162	158–179

Weights at ages 25-59 based on lowest mortality. Weight in pounds according to frame (in indoor clothing weighing 3 lbs., shoes with 1-inch heels.

(Source of basic data: 1979 Build Study, Society of Actuaries and Association of Life Insurance Medical Directors of America, 1980. Table courtesy of Metropolitan Life Insurance Company.)

Once you have determined your best weight, the next step is to factor in your level of physical activity. Follow along with the example given in the table below. Estimate your level of activity as high, medium, or low. A low activity level means that you are mostly sedentary. Playing golf, playing cards, and doing light housework are all examples of low-level physical activities.

A person with a medium activity level is physically active for about ten to fifteen minutes about three times a week. Activities might include swimming, brisk walking, cycling, or playing tennis.

If you are more active than this, you can consider yourself to have a high activity level. (Use your own judgment; these calculations only provide a rough estimate of your needs.) Multiply your best weight in pounds by the number of calories per pound that corresponds to your usual level of activity. The number that results is an estimate of your daily calorie needs.

Now, multiply your total daily calorie needs by your goal for the percent of calories to come from fat. Again, look at the example provided. In the example, the goal is to limit fat to 10 percent of the calories. So, the total calorie intake is multiplied by 0.10. The number that results is the total number of calories per day that will come from fat.

One easy step remains. Since each gram of fat provides nine calories, divide the total number of calories from fat by nine. This number is your goal: the number of grams of fat per day that you will budget. In other words, limit yourself to this number of grams of fat per day. (Note: this section pertains to *adults* who wish to hold their intake of fat to 10 to 15 percent of calories. Fat intakes this low may not be appropriate for young children.)

CALCULATING YOUR FAT GRAM GOAL

STEP ONE: Determine your best weight.
    Example: 130 pounds
STEP TWO: Determine your activity level. Multiply the corresponding value by your best weight.
Low activity = 12 calories per pound
Medium activity = 13 calories per pound
High activity = 14 calories per pound
    Example: If your activity level is medium, 130 pounds x 13 calories per pound = 1,690 calories
STEP THREE: Determine the number of calories per day that will come from fat.
    Example: If your goal is for 10 percent of calories to come from fat: 1,690 calories x 0.10 = 169 calories from fat
STEP FOUR: Determine the number of grams of fat per day that will be your limit.
    Example: 169 calories from fat per day divided by nine calories per gram of fat = 18.7 grams of fat, or about 19 grams of fat. Your limit will be 19 grams of fat per day.

QUICK GUIDE TO FAT GRAM GOALS

For a 1,200-calorie diet:

PERCENT OF CALORIES FROM FAT	FAT LIMIT PER DAY
15 percent	20 grams
10 percent	13 grams

For a 1,500-calorie diet:

PERCENT OF CALORIES FROM FAT	FAT LIMIT PER DAY
15 percent	25 grams
10 percent	17 grams

For an 1,800-calorie diet:

PERCENT OF CALORIES FROM FAT	FAT LIMIT PER DAY
15 percent	30 grams
10 percent	20 grams

For a 2,000-calorie diet:

PERCENT OF CALORIES FROM FAT	FAT LIMIT PER DAY
15 percent	33 grams
10 percent	22 grams

For a 2,300-calorie diet:

PERCENT OF CALORIES FROM FAT | FAT LIMIT PER DAY
15 percent | 38 grams
10 percent | 26 grams

For a 2,500-calorie diet:

PERCENT OF CALORIES FROM FAT | FAT LIMIT PER DAY
15 percent | 42 grams
10 percent | 28 grams

For a 2,800-calorie diet:

PERCENT OF CALORIES FROM FAT | FAT LIMIT PER DAY
15 percent | 47 grams
10 percent | 31 grams

For a 3,000-calorie diet:

PERCENT OF CALORIES FROM FAT | FAT LIMIT PER DAY
15 percent | 50 grams
10 percent | 33 grams

Recommended Resources

There are many excellent general information books available on the subject of vegetarian diets. Those that follow are only a sampling:

Good General Vegetarian Resource Books

Akers, Keith. *A Vegetarian Sourcebook.* Denver, Colo.: Vegetarian Press, 1993.

Melina, Vesanto, R.D., Brenda Davis, R.D., and Victoria Harrison, R.D. *Becoming Vegetarian.* Summertown, Tenn.: The Book Publishing Company, 1996.

Messina, Mark, and Virginia Messina. *The Vegetarian Way.* New York: Harmony Books, 1996.

Low-Fat Vegetarian Books

Attwood, Charles, M.D. *Dr. Attwood's Low-Fat Prescription for Kids.* New York: Viking, 1995.

Barnard, Neal, M.D. *The Power of Your Plate.* Summertown, Tenn.: The Book Publishing Company, 1990.

Barnard, Neal, M.D., with recipes by Jennifer Raymond. *Food for Life: How the New Four Food*

Groups Can Save Your Life. New York: Harmony Books, 1993.

Havala, Suzanne, M.S., R.D., with recipes by Mary Clifford, R.D. *Simple, Lowfat & Vegetarian: Unbelievably Easy Ways to Reduce the Fat in Your Meals*. Baltimore, Md.: Vegetarian Resource Group, 1994.

McDougall, John, M.D. *The McDougall Program*. New York: NAL Penguin, 1990.

Moran, Victoria. *Get the Fat Out! 501 Quick, Easy, and Natural Ways to Cut the Fat in Any Diet*. New York: Harmony Books, 1994.

Ornish, Dean, M.D. *Dr. Dean Ornish's Program for Reversing Heart Disease*. New York: Ballantine Books, 1990.

———. *Eat More, Weigh Less: Dr. Dean Ornish's Life Choice Program for Losing Weight Safely While Eating Abundantly*. New York: HarperCollins, 1993.

Pritikin, Nathan, and Patrick M. McGrady Jr. *The Pritikin Program for Diet and Exercise*. New York: Bantam Books, 1980.

Cookbooks

There are many wonderful vegetarian cookbooks. However, some of those nearest and dearest to my heart for sentimental reasons are not listed here due to the high fat content of many of their recipes. The cookbooks included here consist entirely of very low-fat vegetarian recipes or contain a significant number that fit the bill. Also note that some of the low-fat vegetarian books listed in the preceding sections contain recipes, too.

Hinman, Bobbie. *The Meatless Gourmet: Favorite Recipes from Around the World*. Rocklin, Calif.: Prima Publishing, 1995.

Hinman, Bobbie, and Millie Snyder. *Lean and Luscious and Meatless: Over 350 Delicious, Meat-Free Recipes for Today's Low-Calorie, Low-Cholesterol Lifestyle*. Rocklin, Calif.: Prima Publishing, 1991.

McDougall, Mary. *The McDougall Health-Supporting Cookbook*. Vols. 1 and 2. Clinton, N.J.: New Win Publishing, 1985, 1986.

Rosensweig, Linda. *New Vegetarian Cuisine: 250 Low-Fat Recipes for Superior Health*. Emmaus, Penn.: Rodale Press, 1994.

Wasserman, Debra. *The Low-Fat Jewish Vegetarian Cookbook: Healthy Traditions from Around the World*. Baltimore, Md.: Vegetarian Resource Group, 1994.

Wasserman, Debra, and Reed Mangels, Ph.D., R.D. *Simply Vegan*. Baltimore, Md.: Vegetarian Resource Group, 1991.

Other Publications

Vegetarian Journal, published bimonthly by the Vegetarian Resource Group, P.O. Box 1463, Baltimore, MD 21203, or call 410-366-VEGE.

Vegetarian Times magazine, P.O. Box 570, Oak Park, IL 60303, or call 1-800-435-9610 for subscription orders and information.

Organizations

The Vegetarian Resource Group
P.O. Box 1463
Baltimore, MD 21203
410-366-VEGE

The Vegetarian Resource Group is a national nonprofit vegetarian education organization that publishes the bimonthly *Vegetarian Journal* as well as excellent pamphlets, books, and other materials. Write or call for a catalog of resources available.

Vegetarian Nutrition Dietetic Practice Group
Division of Practice
The American Dietetic Association
216 West Jackson Blvd., Suite 800
Chicago, IL 60606–6995
312–899–0040

The Vegetarian Nutrition Dietetic Practice Group produces a quarterly newsletter, *Issues in Vegetarian Nutrition*, that is available to ADA members as well as to the general public. Call the ADA for subscription information.

Index

About the Author

Suzanne Havala is a registered dietitian and professional nutrition consultant. In addition to working with food companies, nonprofit groups, and other organizations, she writes articles, appears on radio and television, and lectures to professionals and the general public. Among her special areas of interest are health promotion, food trends, and vegetarian diets.

She is the primary author of the American Dietetic Association's position paper on vegetarian diets, and is a founding member and former chairperson of the ADA's Vegetarian Nutrition Dietetic Practice Group. Ms. Havala is also a nutrition adviser for the national nonprofit Vegetarian Resource Group, and she serves on the editorial advisory board of *Vegetarian Times* magazine.

The author is a regular contributor to *Vegetarian Journal* and *Rochester Business Magazine* and has written for *Vegetarian Times* and *Environmental Nutrition Newsletter*. She is frequently quoted in national magazines and newspapers, such as the *New York Times*, *Parade*, *Shape*, *Runner's World*, *New Woman*, *YM*, *Omni*, *Sassy*, *Harper's Bazaar*, and many others, and has appeared on *Good Morn-

ing America and the *Susan Power Show.* Ms. Havala is also the author of *Simple, Lowfat & Vegetarian.*

Ms. Havala is certified as a charter Fellow of the American Dietetic Association. She holds a bachelor of science degree in dietetics from Michigan State University and a master of science degree in human nutrition from Winthrop University, Rock Hill, South Carolina. Based in Charlotte, North Carolina, she has been a vegetarian for over twenty years.

To inquire about her lectures, workshops, or grocery store tours in your area, please send a self-addressed, stamped envelope to:

Suzanne Havala Nutrition Consultants, Inc.
P.O. Box 221383
Charlotte, North Carolina 28222–1383